Excursion Guide
to the Geology of
East Sutherland
and Caithness

Second Edition

Edited by

N. H. Trewin and A. Hurst

DUNEDIN

Published by
Dunedin Academic Press Ltd
Hudson House
8 Albany Street
Edinburgh EH1 3QB
Scotland

www.dunedinacademicpress.co.uk

ISBN 978-1-906716-01-1

© 2009 Aberdeen Geological Society

The first edition of this book was published for
The Geological Society of Aberdeen in 1983
by Scottish Academic Press (ISBN 0 7073 0731 7)

The rights of Nigel Trewin, Andrew Hurst and the named contributors to be identified as authors of this Work has been asserted by them in accordance with sections 77 and 78 of the Copyright, Designs and Patents Act 1988.

All rights reserved.
No part of this publication may be reproduced or transmitted in any form or by any means or stored in any retrieval system of any nature without prior written permission, except for fair dealing under the Copyright, Designs and Patents Act 1988 or in accordance with a licence issued by the Copyright Licensing Society in respect of photocopying or reprographic reproduction. Full acknowledgment as to author, publisher and source must be given. Application for permission for any other use of copyright material should be made in writing to the publisher.

BRITISH LIBRARY CATALOGUING IN PUBLICATION DATA
A catalogue record for this book is available from the British Library

Design and pre-press production by
Makar Publishing Production, Edinburgh

Printed and bound in Poland, produced by Hussar Books

Contents

Acknowledgements and Society information	iv
Editorial Introduction	vii
Geological History of East Sutherland and Caithness – N.H. Trewin	1
Excursion Planner	37

EXCURSIONS

1. The Triassic and Lower Jurassic of Golspie N.H. Trewin	41
2. Bathonian to Oxfordian strata of the Brora Area A. Hurst	48
3. The Upper Jurassic of the Helmsdale Area A.C. McDonald & N.H. Trewin	75
4. The Lower Old Red Sandstone and Helmsdale Granite of the Ousdale Area N.H. Trewin	108
5. The Old Red Sandstone of Caithness N.H. Trewin	116
6. Kildonan gold C.M. Rice	171
References	177

Acknowledgements

The Aberdeen Geological Society is pleased to thank the following for their support of the Society and for generous financial contributions towards the production of this guide; the Petroleum Exploration Society of Great Britain, Senergy through David Harrison, and HRH through Henry Allen.

The sustaining members of Aberdeen Geological Society, HRH, Senergy and BP are also thanked for their support of the annual programme of lectures and other activities of the Society.

Rob Strachan and Ian Alsop are thanked for discussion on basement geology, and making available relevant sections of the Moine excursion guide (currently in press). Phil Gurr for information on the Caithness flagstone industry. Barry Fulton for redrafting many of the figures from the first edition of this work, and creating figures for new excursions. Walter Ritchie for assistance with photography.

The Geological Society of Aberdeen

Membership of the Society is open to all with an interest in geology. A programme of lectures, excursions and social events is organised each year. For further details visit the Society website or write to:

> The Secretary
> Aberdeen Geological Society
> c/o School of Geoscience
> Department of Geology and Petroleum Geology
> Meston Building
> King's College
> Aberdeen AB24 3UE

The Petroleum Exploration Society of Great Britain

The Petroleum Exploration Society of Great Britain (PESGB) was set up in 1964 as a non-profit making organisation and is a registered charity. It has a membership of 5,500 individual members and over 80 Sustaining (company) members. The object of the Society is to promote, for the public benefit, education in the scientific and technical aspects of petroleum exploration.

Monthly meetings are held in London and Aberdeen. Distinguished represenwwtatives from industry and academia are invited to present a talk on a wide variety of topics associated with petroleum exploration.

In addition to organising the biennial PETEX Conference and Exhibition, the PESGB holds other conferences, seminars and courses, some in conjunction with the Society's Special Interest Groups (SIGs), and others jointly with other similar Societies. Regular events include: Geophysics seminars, Data Management conferences and courses; North Sea courses; DEVEX Conference & Exhibition; HGS/PESGB Africa Conference; PROSPEX Fair and the Petroleum Geology Conferences (PGC).

The PESGB also sponsors and supports other activities relevant to its objectives. In 2008 it introduced a Scholarship Award Scheme for MSc students studying an Earth Science degree course.

The Society runs field trips to areas of petroleum and geological interest, both in the UK and abroad, and organises occasional core workshops.

The Society produces a monthly Newsletter and an annual Membership Directory, which are distributed to all members. Recent additional benefits to members include: each new edition of The North Sea Map, the Millennium Atlas DVD, and DVDs of the PGC IV, V and VI Proceedings.

There are four classes of members – Active, Associate, Student and Sustaining. The annual membership subscription is £25.

For further information about the PESGB, please contact the PESGB Office, 5th Floor, 9 Berkeley Street, London W1J 8DW, Tel: +44 (0)20 7408 2000, Fax: +44 (0)20 7408 2050, email: pesgb@pesgb.org.uk; web: www.pesgb.org.uk.

Addresses of authors

A. Hurst, C.M. Rice and N.H. Trewin
 Department of Geology and Petroleum Geology
 School of Geosciences
 University of Aberdeen
 Meston Building
 King's College
 Aberdeen
 AB24 3UE

A.C. McDonald
 Roxar Oslo
 Hoffsveien 1C
 6th Floor,
 PO Box 165
 0275 Oslo
 Norway

Editorial Introduction

The Jurassic rocks of the Brora–Helmsdale area form a coastal strip from Golspie in the south to Helmsdale in the north and are bounded on the landward side by the Helmsdale Fault. To the west of the fault the country rocks consist of Moinian metamorphics intruded by the late Caledonian Helmsdale Granite, and these rocks are overlain unconformably by the Old Red Sandstone of Devonian age.

This area has an interesting and varied geological history and has been the focus of considerable geological research from the early work of Murchison (1827) right up to the present day. Early interest was stimulated by the presence of Jurassic coal at Brora, which was mined from as early as 1598 until 1974. Bricks were made from Jurassic clays at Brora but the site of the brickpits has now been landscaped. A brief 'gold rush' to Kildonan took place in 1868, and gold can still be panned at Baille an Or.

Numerous geological field parties visit the area each year, usually to look at the spectacular Upper Jurassic 'Boulder Beds' which were deposited in deep water on the downthrow side of the Helmsdale Fault at a time when the fault was active. Other geologists are drawn to the area to see the lower part of the Jurassic succession, which has affinities to the rocks of the Beatrice Oilfield whose production platforms can be seen some 12 miles off Helmsdale on a clear day. To the north of Helmsdale lie the famous Old Red Sandstone deposits of Caithness, dominated by cyclic lake deposits with spectacular faunas of fossil fish.

The purpose of this volume is twofold; the first being to introduce the reader to the geology of the area by means of a chapter on the geological history of the region, and the second to provide a series of excursions to illustrate the geology. Some previous knowledge of geology is generally assumed, but all excursions are intended to provide interest for the amateur geologist, student and professional geologist. A checklist of major topics covered in the excursions, together with details of restricted access to some localities, is given in the Excursion Planner section of this guide. A geological map of the area is included as Figure 1 and the symbols used on this map have been utilised throughout the volume as far as possible.

A range of excursions is provided which would occupy a party for a week or more. There are good hotels, guest houses, caravans, youth hostels and camping sites in the area, details of which can be obtained from the Tourist Information outlets. The towns of Golspie, Brora and Helmsdale are convenient for the Jurassic outcrops, and Wick and Thurso for the Devonian, but all excursions are within day-trip range of Helmsdale. Mention must be made of the excellent geological exhibition at the Orcadian Stone Company in Golspie. The Timespan Heritage Centre exhibition beside the old bridge in Helmsdale has good exhibits relating to local history and the history of the gold rush of 1868.

In following the enclosed excursions, always heed the Country Access Code by sensible parking, closing gates and avoiding growing crops and lambing ewes. Also follow the Scottish Fossil Code published by Scottish Natural Heritage. Seek permission to visit exposures in farmland and quarries, and take extreme care crossing the railway line. Do not hammer needlessly at any outcrops, or excavate exposures in a search for fossils; others will come after you wishing to see and photograph outcrops. Specimens can usually be obtained from loose material. In particular, do not hammer or remove bedrock at any Sites of Special Scientific Interest; SSSI – these are noted in the text. Never ignore personal safety, particularly in cliff and moorland areas. The weather in the area can be severe and change very quickly.

Since the publication of the first edition of this guide in 1993 there have been a number of changes in the area affecting access and exposure. Hence some of the original localities have been omitted and others added.

We hope that some of the different views and emphasis provided in the excursions will prove stimulating for the reader. Individual authors are responsible for the scientific content of each excursion, and excursions should be referenced under the author's name.

Nigel H. Trewin, Andrew Hurst
April 2009

Geological history of East Sutherland and Caithness

N.H. Trewin

Contents
1. **Metamorphic Basement** *4*
2. **Helmsdale Granite** *6*
3. **Devonian** *7*
4. **Post-Caledonian Igneous activity** *15*
5. **The Inner Moray Firth Basin** *15*
6. **Permo-Triassic** *17*
7. **Jurassic** *19*
8. **Cretaceous** *28*
9. **Tertiary** *29*
10. **Quaternary** *29*
11. **Economic Geology** *31*

Preface

This excursion guide provides the reader with a series of excursions that can be accomplished using Helmsdale, Brora or Golspie as a base in the south of the area and Thurso or Wick in the north. Whilst the aim of the guide is to provide a geological variety to the excursions, it has to be recognised that the main claim to fame of the Jurassic rocks of the area is the spectacular development of the Kimmeridgian Helmsdale Boulder Beds which were deposited on the downthrow side of the Helmsdale Fault during a phase of fault activity. Every year numerous geologists visit the area to examine these rocks, and the rest of the coastal strip of Triassic-Jurassic rocks exposed on the shore and in river valleys. Excursions are also included to cover the other main features of local geology including the Moine basement, the Helmsdale Granite and the Lower Old Red Sandstone (ORS). In Caithness the classic Middle ORS cyclic lacustrine facies, the lake margin unconformities, and the world-famous fossil fish locality at Achanarras Quarry are the main highlights. A selection of other localities has been chosen to illustrate the main features of the ORS of this part of the Orcadian Basin.

The area has been exploited for its geological riches from early times. The Brora Coal was extracted from at least as early 1598 and worked intermittently up to 1974 and the workings abandoned the following year. In early times the coal was used to produce salt by evaporation of sea water, and latterly for domestic consumption and to fire the kilns of the Brora brickworks, which used the Jurassic Brora Brick Clay. The Jurassic sandstones provided the bulk of the local building freestone from quarries at Strathsteven and Clynelish. The discovery of alluvial gold in 1868 at Kildonan resulted in a minor goldrush, and it is still possible to pan small quantities of gold (Exc. 6). With the exploration of the North Sea for hydrocarbons came the discovery of the Beatrice field in 1976, whose production installations can be seen from Helmsdale on a clear day. The field produces oil from sandstones of Middle and Lower Jurassic age which are similar to those of the Brora area.

Several well-known geologists have been associated with the area. Murchison (1827) surveyed the Brora coalfield and Hugh Miller, famous for his work on the fossil fish of the Old Red Sandstone, also studied the area (Miller, 1854, 1859). Judd (1873) provided a detailed account of the Mesozoic succession; his lengthy paper includes much observation of features no longer visible due to poor exposure, modifications to the A9, and filling of quarries. Read *et al.* (1925) described the general geology of the area in the Golspie Memoir of the Geological Survey, to

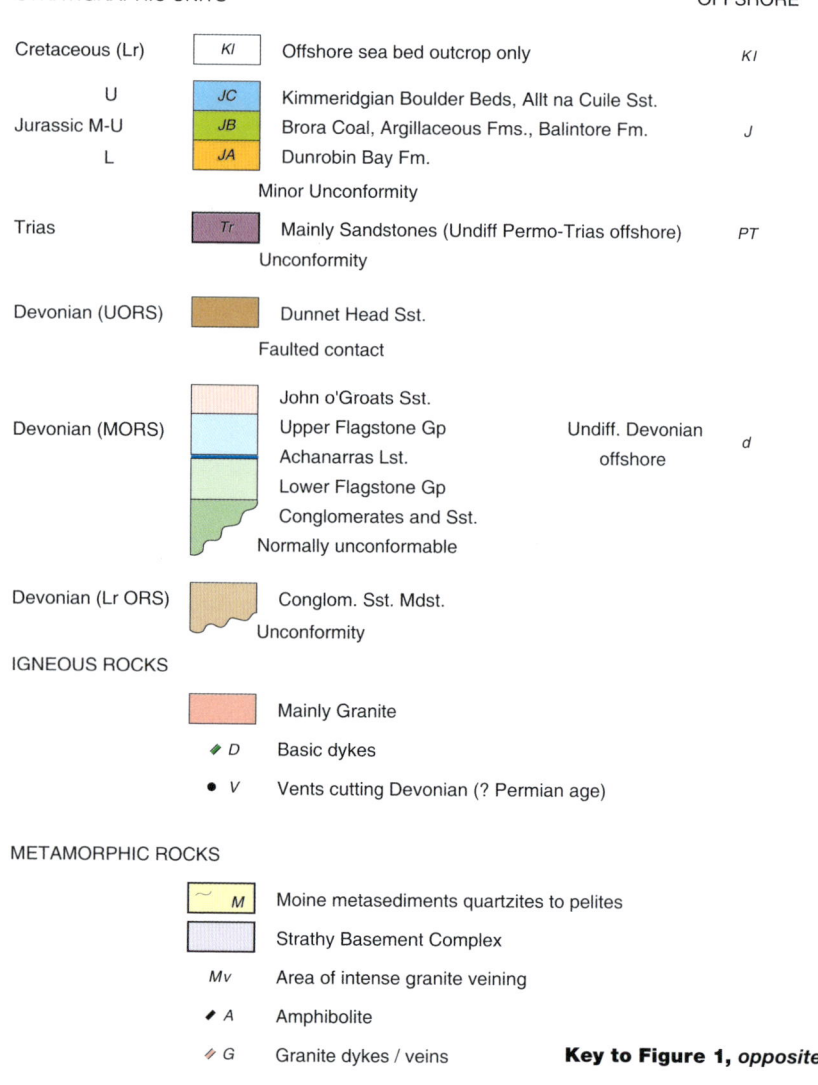

Key to Figure 1, *opposite*.

which Lee contributed valuable detail on the Jurassic. The classic contribution on the Helmsdale Boulder Beds by Bailey and Weir (1932) is a masterpiece of careful observation and logical deduction. More recent authors have added detail, but Bailey and Weir's recognition of the effects of submarine faulting contemporary with deposition remains the cornerstone of subsequent work.

This geological history is arranged as far as possible in chronological order, but some events certainly overlap in time. Inevitably the account that follows is the author's personal interpretation, and reference is made to the literature to enable the reader to follow up points of controversy or alternative interpretations.

A general geological map (Fig. 1) and a stratigraphic summary (Fig. 2) are provided as a reference framework for the reader. Further details of the geological history are included in the introductions to the various excursions. The following account is brief and concentrates on features that are demonstrated by the excursions. More detail, particularly of the tectonic, metamorphic and igneous events, can be found in Trewin (2002) and Macdonald and Fettes (2007).

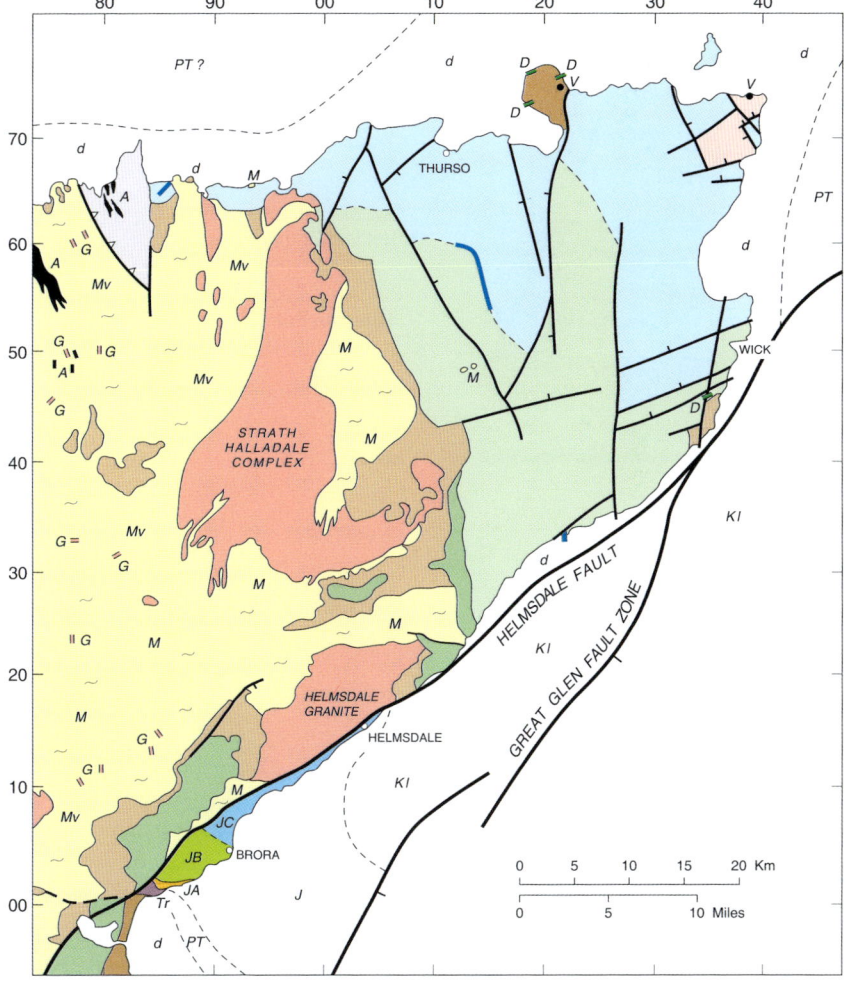

1 Sketch map of the geology of East Sutherland and Caithness.

4 Geological History

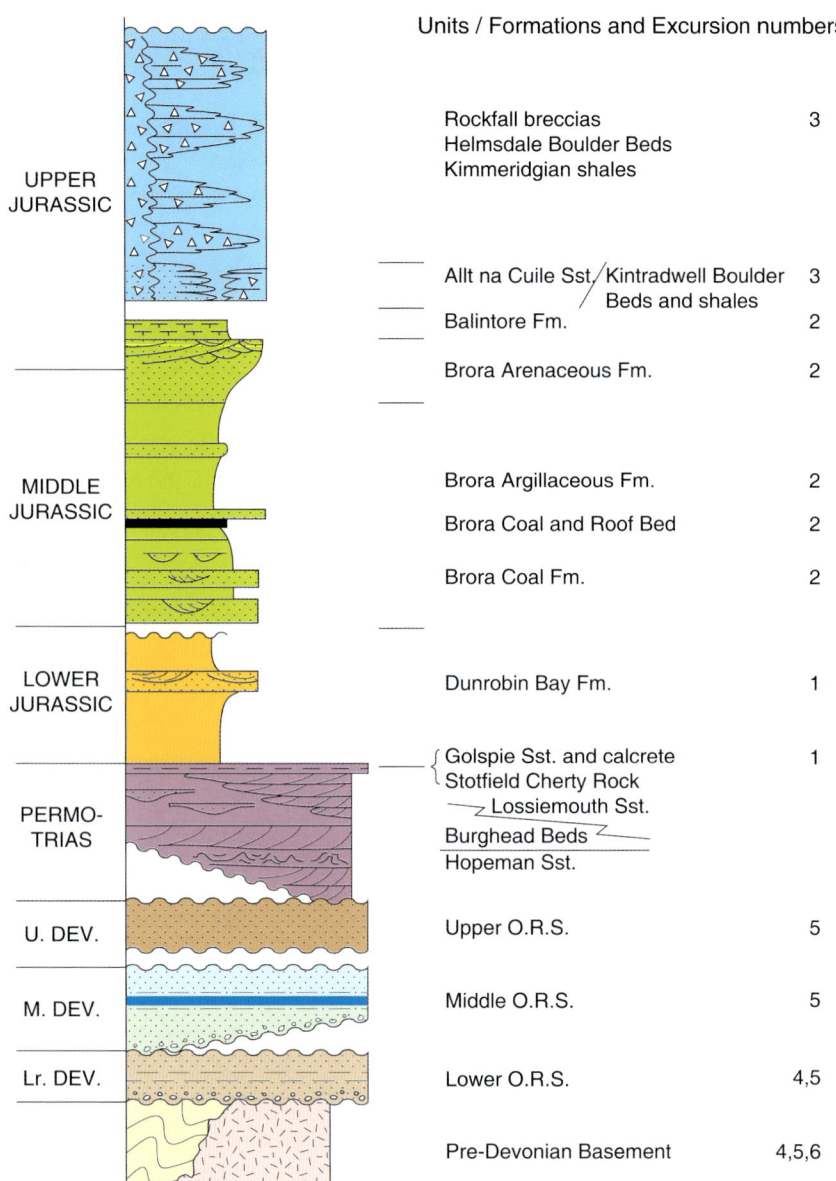

2 Basic stratigraphic framework and relevant excursions.

1. Metamorphic basement

It is not the intention to deal in any detail with the history of the basement in this area for the simple reason that the guide is mainly intended for those interested in the cover sequence. Furthermore, an excellent new guide to Moine geology (Strachan *et al.*, 2009a in press) that updates the previous edition (Allison *et al.*, 1988) includes a geological summary (Strachan *et al.*, 2009b in press), and Excursion 13 in that guide (Strachan *et al.*, 2009c in press) covers the North Sutherland coast in some detail.

From c. 400	Deposition of Lower Old Red Sandstone starting in Emsian, Initiation of Orcadian Basins. Uplift and erosion to expose Helmsdale Granite.
c. 420	Intrusion of Helmsdale Granite.
435 - 425	Scandian metamorphism deformation and nappe formation, ending with Moine Thrust movements and intrusion of undeformed Strath Halladale granite complex.
c. 470 - 440	Grampian metamorphic event, peak in mid Ordovician and including migmatites in East Sutherland. Inclusion of basement slices in Moine.
c. 820 - 870	Knoydartian orogeny seen on west coast of Scotland. Polyphase metamorphism and granite intrusion not proven in E. Sutherland, but some pre-Grampian event probable.
1000 - 900	Deposition of Moine sediments, mainly sandstones and shales, on metamorphic basement.

3 Sequence of events in pre-Devonian basement. Summarised from Strachan *et al.* (2002) and Trewin and Rollin (2002).

The guide can be obtained from the geological societies of Edinburgh and Glasgow. An overall summary of the Moine geology is given by Harris and Johnson (1991) and Strachan *et al.* (2002), and the igneous rocks are covered by Brown (1991). A summary of events affecting the basement is presented in Figure 3.

In this guide, Moine basement is seen on Excursion 6 in Strath Helmsdale at Kildonan, and beneath the sub-Devonian unconformity at Dirlot (Exc. 5) and at Red Point, Port Skerra and Strathy (Exc. 5). The pre-Devonian basement consists mainly of Moine metasediments, with an inclusion of gneiss (Strathy Complex) which is possibly part of the basement to the Moine (Strachan *et al.*, 2002; Macdonald and Fettes, 2007). These rocks form part of the Naver (=Swordly) Nappe which rests on the Swordly Thrust. This nappe is the highest in a pile of nappes which moved from SE to NW and rest ultimately on the Moine Thrust. Much of the Naver Nappe is highly migmatised and is intruded by the East Sutherland (=Strath Halladale) migmatitic complex which has been dated at 461 ± 13 and 467 ± 10 Ma on U/Pb ages from zircon rims, showing it to be part of the Ordovician age Grampian tectonothermal event (Kinny *et al.*, 1999).

Basement to the Moine

The oldest rocks in the area form the Strathy Complex which is thought to be a slice of basement gneiss that was brought to its present position along the Swordly

Thrust during Grampian age deformation. Inliers of basement in the Moine have been called 'Lewisian', and have provided age dates consistent with the Lewisian. The rocks of the Strathy Complex comprise a variety of gneisses and granulites with trondhjemitic sheets and pegmatites (Strachan et al., 2009c in press). The gneisses include siliceous grey gneiss, quartz-magnetite gneiss, garnet–ortho-amphibole gneiss and hornblende gneiss. In addition, calc-gneisses contain scapolite–diopside–spinel assemblages. Cross-cutting amphibolite dykes and post-orogenic microgranites are also present.

In the gneisses, an early metamorphic assemblage is characterised by garnet, ortho-amphibole and staurolite in a quartz-oligoclase granoblastic matrix. This assemblage is postdated by biotite, sillimanite and blebby quartz.

Moine

The rocks in the area of the guide form part of the Naver (Swordly) Nappe. The sediments were originally pelitic with minor psammite, but psammitic rocks are more frequent in the east. The time of deposition is dated at 1000–900 Ma (Fig. 3). The Moine sediments further to the west were deformed by the Knoydartian event around 820–870 Ma, but this event has not been recognised in the area of this guide; evidence may have been largely destroyed by the Grampian event in the Ordovician. The Moine displays an early suite of tholeiitic intrusions now represented by amphibolites, and polyphase deformation has affected these rocks. In the Grampian event upper amphibolite facies metamorphism was associated with two phases of isoclinal folding, and migmatisation took place during the second phase to form the Strath Halladale migmatites. The migmatitic banding was later deformed by tight to isoclinal NW and SE plunging folds, and then intruded by foliated granite and pegmatite sheets. The resulting rocks are characterised by the 'Kirtomy assemblage' of migmatitic pelitic gneisses and granites described in the Moine field guide by Strachan et al. (2009c in press).

Scandian deformation resulted in mainly upright, tight SE and NW plunging folds which refolded the previous set. Movement on the Swordly Thrust produced ductile shear zones. Late Scandian events produced monoformal and brittle conjugate folds, with widespread retrogression from the upper amphibolite facies of the peak of metamorphism. Granites intruded at the end of the Scandian event are undeformed and form the Strath Halladale granitic complex. Examples of late granite veins are seen at Portskerra (Exc. 5).

2. Helmsdale Granite

The Helmsdale Granite (Fig. 1) is a representative of the Newer Granites in the sense of Brown (1991); it appears to be emplaced along the line of the Helmsdale Fault, which is probably a Caledonian structure. The suite of Newer Granites is calc-alkaline in character and compares chemically with modern magmatic arc products generated by subduction. It is probable that the Helmsdale Granite is related to northward Silurian to early Devonian subduction.

The Helmsdale Granite consists of two phases. An outer zone 1–3 km wide consists of coarse pink, porphyritic adamellite with large pink K-feldspar phenocrysts, and the inner zone is a fine-grained pink-grey adamellite. The two phases appear to have been arranged in a concentric manner and the Helmsdale Fault effectively cuts the intrusion in half. The intrusive contact with the Moine is sharp and steep with little evidence of hornfelsing; presumably the magma was permissively intruded into cold country rocks at a high structural level. The contact between the two intrusive phases is gradational over distances of a few metres to a few hundred metres. Each type of granite has been recorded cutting the other (Tweedie, 1981), and although the fine-grained phase was intruded last, the porphyritic phase was apparently still sufficiently mobile to back-vein the later phase.

Both phases have a similar mineralogy with roughly equal amounts of quartz, plagioclase and K-feldspar. The plagioclase is albite-oligoclase and the K-feldspar mostly orthoclase with some microcline. Both feldspar types are frequently zoned and have zonal sericitic alteration. Biotite makes up to 5% of the rock and accessories include zircon, apatite, magnetite and altered titanite (sphene) (Tweedie, 1979).

The granite was deeply weathered to a grus prior to deposition of the Lower Old Red Sandstone (Emsian), and the pink K-feldspar phenocrysts are a common constituent of the overlying Ousdale Arkose; indeed, it can be difficult to tell the arkose from the igneous rock in the field.

The intrusion has been dated at $c.420$ Ma by the U-Pb method on zircon (Pidgeon and Aftalion, 1978). The overlying sediments are dated at about 400–390 Ma (Emsian), and thus there was an interval of some 20–30 Ma available for uplift, erosion and unroofing of the granite following cooling. The present day erosion surface is probably similar to the Devonian level.

3. Devonian

Introduction

The Devonian strata of northern Scotland were immortalised by Hugh Miller in his work 'The Old Red Sandstone' (1841). Hugh Miller's Cottage and Miller House at Cromarty, on the Black Isle, are now a museum of the National Trust for Scotland and well worth a visit. Thick deposits of Old Red Sandstone are present in onshore areas bordering the Moray Firth and consist of a great variety of fluvial, lacustrine and subordinate aeolian strata deposited in the SW portion of the Orcadian Basin. This large inland basin extended north through Orkney to Shetland and as far east as western Norway. In practice, this area included a large number of smaller basins, largely formed by movement on bounding fault systems (e.g. Golspie and Badbea Basins of Dec (1992)). These basins, and the highs that separated them, formed part of a large strike-slip system between Greenland and N. America to the west and Europe to the east (Ziegler, 1982). The Orcadian Basin contains deposits of non-marine origin, with only the faintest sniff of a marine influence at one horizon in the Eday Flagstones of Orkney (Marshall *et al.*, 1996). Marine conditions existed in SW England with a shoreline roughly E–W into Europe, but an embayment of the Middle Devonian sea possibly extended north in the position of the present North Sea as far as the Auk and Argyll oilfields some 200 km SE of Aberdeen. Marine environments also occur to the east in the Baltic (particularly Estonia) (Marshall *et al.*, 2007) and there are strong similarities between fossil fish

8 Geological History

	Brora Outlier		Caithness		
FAMENNIAN / FRASNIAN	NOT EXPOSED		DUNNET HEAD SANDSTONE GROUP — — — BASE NOT SEEN — — —		
GIVETIAN			JOHN O' GROATS SANDSTONE GROUP		
GIVETIAN			UPPER CAITHNESS FLAGSTONE GROUP 1500 m +	MEY SUB-GROUP 553 m	
				HAM-SKARFSKERRY SUB-GROUP 750 m	
				LATHERON SUB-GROUP 175 m	
					SPITAL SUB-GROUP
EIFELIAN			LOWER CAITHNESS FLAGSTONE GROUP 2350 m	ACHANARRAS LIMESTONE MEMBER	
				ROBBERY HEAD SUB-GROUP 155 m	
				LYBSTER SUB-GROUP 870 m	
				HILLHEAD RED BED SUB-GROUP 160 m	
	COL-BHEIN FORMATION	Flaggy sandstone 260 m +		BERRIEDALE FLAGSTONE FORMATION	CLYTH SUB-GROUP 1150 m
	SMEORAIL FORMATION	Conglomeratic and pebbly sandstone		BERRIEDALE Sst. FM.	(= HELMAN HEAD BEDS)
LOWER OLD RED SANDSTONE ? SIEGENIAN AND EMSIAN	Period of folding, locally producing marked angular unconformity		BARREN OR BASEMENT GROUP c. 300 m (= SARCLET GROUP) 437 m	BADBEA BRECCIA	ELLEN'S GOE CONG.
				Angular unconformity in south and west Caithness	ULBSTER/IRES GEO SANDSTONE FM. 107 m
	GLEN LOTH FORMATION	Mudstone and fine grained sandstone 600-700 m		OUSDALE BRAEMORE, etc MUDSTONES	ULBSTER/IRES GEO MUDSTONE FM. 172 m
	BEN LUNDIE FORMATION	Basal breccia-conglomerate and arkose up to 200 m		OUSDALE ARKOSE	ULBSTER/IRES GEO Sst. FM. 85 m
					SARCLET CONG. FM. 70 m
	BASEMENT			HELMSDALE GRANITE	Base not seen
					METAMORPHIC BASEMENT

4 Stratigraphic nomenclature for the Devonian in eastern and southern Caithness and the Brora region of Sutherland. Modified from Trewin and Thirlwall (2002).

faunas from Estonia and Caithness (Newman and Trewin, 2008). Rivers and lakes in the Orcadian basin may have ultimately drained to the sea in times of overflow (Trewin, 1986). Areas of compression and areas of extension were created at different times within the Orcadian Basin. There is evidence of extensional tectonics controlling sediment deposition, as well as unconformities which record periods of intra-Devonian faulting, uplift and erosion.

Some authors (Norton *et al.*, 1987, Coward and Enfield, 1987) consider that Devonian sedimentation in the Orcadian area was controlled by 'extensional collapse' of Caledonian crust and that sedimentation took place in half-grabens. The unconformities recognised are then ascribed to footwall uplift and erosion in a generally extensional regime. Those who consider Devonian strike-slip to have been important prefer to interpret the evidence for extension in terms of the presence of transtensional basins, with sediment derived by erosion of transpressional highs (Trewin, 1989). The fact that erosional events appear to be roughly synchronous over wide areas may indicate that there were specific compressive periods associated with strike-slip between Greenland and N. America, but that extensional collapse was superimposed on this system. Work by Underhill and Brodie (1993) using seismic data from the Easter Ross area identifies a period of Lower Devonian extensional faulting which was probably related to rifting, followed by a period of Middle to Upper Devonian regional subsidence.

In northern Scotland the Devonian succession (Fig. 4) is divided into Lower, Middle and Upper Old Red Sandstone (ORS). These divisions approximate to Lower, Middle and Upper Devonian time. The stratigraphic nomenclature is complex, with a plethora of names used in different parts of the basin. It is apparent from recent work on the fish faunas that stratigraphic revision is required, and the British Geological Survey (BGS) (2005) have published revised nomenclature for the area near Dounreay (Fig. 5) firmly based on fish faunas. The reader is warned that the stratigraphic terminology is currently confusing, since differing schemes and names appear on BGS maps of the area depending on date of publication. The large thicknesses of the Flagstone Groups noted in Figure 4 are probably excessive due to measurement of total aggregate thicknesses, and lack of appreciation of lateral facies variation. Geophysical gravity data indicates that the ORS succession at any one place in Caithness seldom exceeds 2.5 km. Current work by BGS, incorporating new faunal evidence will lead to better definition of the stratigraphy of the Flagstone Groups in Caithness, and revision of geological maps.

Near the basin margins the three units of the ORS (Fig. 1) are separated by unconformities, as present in the Ousdale area (Exc. 4), representing lengthy periods of uplift and erosion. Away from the margin there is apparent conformity between Lower and Middle ORS, as at Sarclet, but a major change in clast derivation is seen. Conformity may also exist between Middle and Upper ORS (Rogers *et al.* (1989), but all contacts are faulted in Caithness. However, a clear unconformity exists on Hoy, Orkney.

Lower Old Red Sandstone

The Lower ORS in the Helmsdale area is quite variable in character and thickness, and was deposited on the irregular eroded topography of the post-Caledonian land surface. A long period of erosion took place prior to Lower ORS deposition, since the Lower ORS at Ousdale, which is dated on spores as early Emsian ($c.$ 390 Ma) (Collins and Donovan, 1977), rests on eroded Helmsdale Granite which is dated at $c.$ 420 Ma (Pidgeon and Aftalion, 1978). It is probable that at least 3 km of rock cover was eroded over this period to unroof the Helmsdale Granite, and Moine metamorphic rocks.

The basal deposits of the Lower ORS generally closely reflect the underlying basement geology from which the clasts were derived. Thus, at Ousdale an arkose (Ousdale Arkose, Exc. 4) is present, consisting almost entirely of material derived by

Geological History

Age	Groups / subgroup	Formation / Member	Vertebrate Biostratigraphical Zones		
			Osteolepid Zone	Coccosteid Zone	Dipnoan Zone
GIVETIAN / EIFELIAN	UPPER CAITHNESS FLAGSTONE SUBGROUP	MEY FLAGSTONE FORMATION	*Thurius pholidotus*	*Millerosteus minor*	*Dipterus valenciennesi*
				No arthrodires found to date	
EIFELIAN		SPITAL FLAGSTONE FORMATION	*Gyroptychius milleri*	*Dickosteus threipalandi*	
EIFELIAN	LOWER CAITHNESS FLAGSTONE SUBGROUP	Achanarras Fish Bed Member	*Osteolepis macrolepidotus*	*Coccosteus cuspidatus*	*Pinnalongus saxoni*
		LYBSTER FLAGSTONE FORMATION	*Thursius macrolepidotus*	No arthrodires found to date	No dipnoans found to date
EMSIAN	SARCLET GROUP		No biostratigraphically useful fish fossils		

Age	Groups / subgroup	Formation / Member	Vertebrate Biostratigraphical Zones		
			Osteolepid Zone	Coccosteid Zone	Dipnoan Zone
GIVETIAN	UPPER CAITHNESS FLAGSTONE SUBGROUP	CROSSKIRK BAY FORMATION	↑ *Gyroptychius milleri*	↑ *Dickosteus threipalandi*	↑ *Dipterus valenciennesi*
	---- ? ----	DOUNREAY SHORE FORMATION	No osteolepis found to date	No arthrodires found to date	
EIFELIAN	LOWER CAITHNESS FLAGSTONE SUBGROUP	SANDSIDE BAY FORMATION	*Thursius macrolepidotus*	*Coccosteus cuspidatus*	*Pinnalongus saxoni*
		BIGHOUSE FORMATION		No arthrodires found to date	No dipnoans found to date
EMSIAN	SARCLET GROUP		No biostratigraphically useful fish fossils		

5 Stratigraphic nomenclature for the Devonian in NW Caithness and the adjacent part of Sutherland showing correlation and fish faunas on either side of the Bridge of Forss Fault. (From British Geological Survey, 2005; Newman and den Blaauwen, 2008.)

weathering of the Helmsdale Granite. To the north at Sarclet metamorphic clasts are much in evidence in conglomerates, although the unconformity is not seen. To the south in the Golspie Basin (Brora outlier), arkosic conglomerate dominates where Helmsdale Granite underlies the ORS, and metamorphic clasts in areas underlain by Moine (Read et al., 1925). The conglomerates of the Golspie Basin were deposited on alluvial fans dominated by debris-flow processes (Dec, 1992).

Deep weathering in Devonian times resulted in granular disintegration of the Helmsdale Granite and the unconformity between arkose and granite is not always clearly defined (Exc. 4). The basal conglomerate and arkose are of variable thickness (up to 200 m) and were deposited on the irregular land surface by alluvial fans and streams, and also by flash floods which followed periodic storms in an otherwise semi-arid climate. Above the coarse basal deposits red mudstones with brown-red sandstones are present. The Ousdale Mudstones are typical of this part of the succession and formed in an alluvial plain environment which was crossed by river channels. Streams were probably ephemeral and flash floods occasionally spread coarse material over the dried mud surfaces. Some permanent water must have been present since a fauna of arthropods left evidence of their activity in the form of trace fossils, and plant debris is present in some of the channel sandstones.

Middle Old Red Sandstone

In the Brora area and north of Helmsdale at Badbea (Exc. 5) the Middle ORS basal conglomerate rests unconformably on Lower ORS and also oversteps Lower ORS to rest on basement (Mykura 1991). Further north at Sarclet the Ellens Goe Conglomerate at the base of the Middle ORS rests with apparent conformity on Lower ORS, but exposures are limited in that area. In general, it appears that a strong phase of folding, faulting, uplift and erosion affected the Moray Firth area, as similar unconformable relationships are seen in many localities to the south of the Moray Firth (Trewin and Thirlwall, 2002), for example at Pennan and Gamrie (Trewin, 1987; Trewin and Kneller, 1987). However, Rogers et al. (1989) consider that unconformities, when present, are only of local significance and represent the result of rejuvenation of extensional faults.

To the south of Berriedale the Middle ORS consists mainly of sandstones of broadly fluvial origin, and conglomerates deposited by alluvial fans, the major exception being the lacustrine interlude represented by a thin sequence of laminated siltstones with fish-bearing calcareous concretions, which fired the enthusiasm of Hugh Miller in the last century. These nodule beds at Gamrie, Tynet Burn, Edderton, Eathie and Cromarty (Fig. 6) contain the Achanarras fauna of fish and form an extensive marker horizon of Eifelian age in the Orcadian Basin.

To the north of Berriedale the thick (up to 3.9 km aggregate thickness) Caithness Flagstone Groups are found. These rocks are predominantly grey or green in colour, usually dolomitic, and form a series of cyclic deposits recording deposition in a lake of fluctuating depth and extent (Donovan, 1980).

During deep lake periods the dark organic-rich laminated siltstones of the fish beds were deposited in anoxic conditions. The fish carcasses drifted out into the lake and sank in deep water to be preserved due to a lack of scavengers, and the

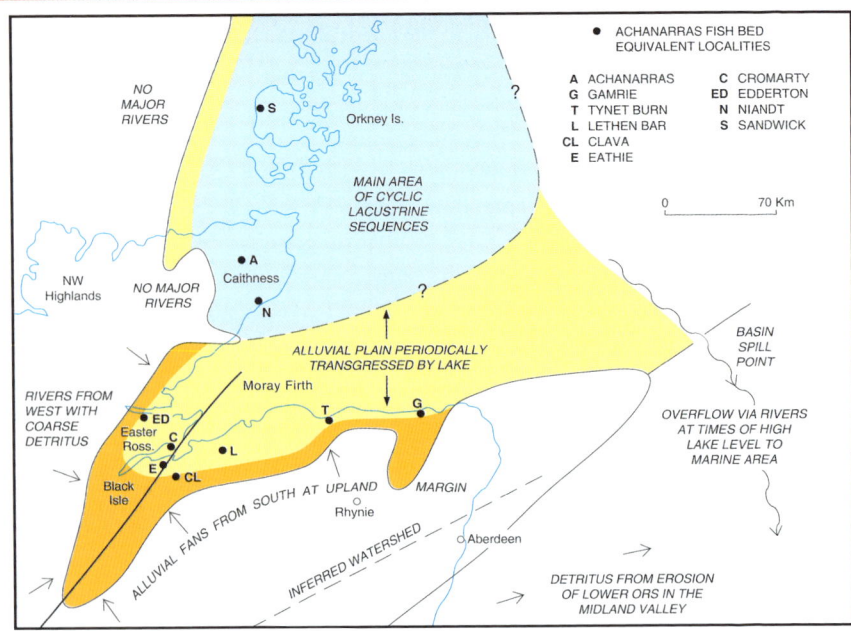

6 Middle ORS palaeogeography of north-east Scotland reconstructed with 30 km post-Devonian dextral shift on the Great Glen Fault. Modified from Mykura (1991) and Hamilton and Trewin (1988).

presence of favourable chemical conditions (Trewin, 1986). Most of the fish probably lived in the shallower areas of the lake, where conditions were not suitable for their fossilisation.

As the lake became shallower, more coarsely laminated silts and fine sands were deposited. Lenticular subaqueous shrinkage cracks (Donovan and Foster, 1972) are common, and probably reflect changing salinity in the lake, as is demonstrated by the isotopic ratios of carbon and oxygen, and carbon/sulphur ratios (Hamilton and Trewin, 1988; Duncan and Hamilton, 1988). In the shallow lake phases, fine-grained sandstones with ripple lamination are common, and drying out of the lake is marked by horizons of polygonal desiccation cracks. Evidence of evaporitic conditions is provided by pseudomorphs after anhydrite and halite, and in the North Sea (Well 9/16-3) an anhydrite layer has been recorded representing local hypersaline conditions (Duncan and Buxton, 1995).

The repeated cycles of the Caithness Flagstone Groups are thought to have been climatically controlled. The cycles represent changes from cooler/wetter to warmer/dryer climatic conditions that caused the Orcadian lake to expand and contract (Hamilton and Trewin, 1988). There is agreement that the cycles record Milankovitch periodicities caused by regular changes in the inclinations of the axis of rotation of the earth and in its orbit. The cycles have been assigned to $c.20,000$ years (Precession) $c.40,000$ years (Obliquity) and $c.100,000$ years (Eccentricity) periodicities by various authors (Hamilton and Trewin, 1988; Astin, 1990; Kelly, 1992; Marshall *et al.*, 1996), but authors have not agreed on the duration of the cycles. The problem arises with the choice of depositional rate, and application of the Devonian timescale to the Middle ORS of the area to give an accurate duration for any part of the succession. Recently Andrews (2008) has reviewed all the factors on the basis of the most recent data and concludes that Precession exerted the main control, modulated by Eccentricity. Calculated Devonian values for Precession cycles are 16.7 and 19.7 Ka.

The Achanarras fish bed (Exc. 5) marks a particularly extensive and deep lake phase, and lies at the base of the thickest cycle in the sequence. The fish-bearing part of the sequence was probably deposited over a period of more than 4,000 years, and it contains a remarkable record of changing fish faunas in the Orcadian lake.

Fifteen genera of fish are present at Achanarras including *Dipterus*, *Coccosteus*, *Pterichthyodes*, *Palaeospondylus* and *Mesacanthu*s as common representatives, and rarer *Glyptolepis*, *Osteolepis*, *Diplacanthus*, *Cheiracanthus*, *Rhamphodopsis*, *Homosteus* and *Cheirolepis*. Further details of the fauna are given in Trewin (1986), and in Excursion 5. A full variety of omnivores (*Dipterus*), scavengers (*Pterichthyodes*) and predators (*Coccosteus* and *Glyptolepis*) are present, representing a food chain with ultimate reliance on phytoplankton. Many of the fish are exceptionally well preserved, and probably died during mass-mortality events.

At the greatest extent of the lake during deposition of the Achanarras fish bed it is probable that the lake overflowed via rivers to the sea to the SE (Fig. 6). It is thought that the fish originally migrated into the inland lake system from the sea (Trewin, 1986), since relatives of most of the fish also occur in marine environments, and in other, widely separated Devonian continental areas.

Whilst lake sediments dominate the thick basin-centre sequences, the basin margin deposits include deposits of fluvial floodplains, lake shoreline sands and small aeolian dunes (e.g. Sandside Bay, Excursion 5). In places, rocky shorelines against Moine basement existed (e.g. Dirlot Castle, and Red Point, Excursion 5). The general lake margin environments are illustrated in Figure 7.

By the end of the Middle Devonian a thick pile of cyclic lacustrine sediments had accumulated in central Caithness with the basal ORS buried to depths of 3 km or more. At the basin margins less subsidence had taken place. Organic matter in fish beds in the deeply buried section would have been generating oil, the relics of which now occur as bituminous residues in fish beds and sandstones. Details of the organic geochemistry are given by Duncan and Hamilton (1988).

Upper Old Red Sandstone

The Upper Old Red Sandstone of Hoy in Orkney has a lava at the base and rests unconformably on Middle ORS flagstones. Wilson *et al.* (1935) and Mykura (1976) considered the flagstones below the unconformity to be Upper Stromness Flagstones, but they are now thought to belong to the Rousay Flagstone Formation. The North Scapa Fault cuts the Middle ORS, but became inactive prior to Upper ORS deposition (Wilson *et al.*, 1935). It certainly appears that the majority of the Givetian is not represented. In Caithness the Upper ORS Dunnet Head Sandstone is faulted against Middle ORS. Unconformity is present to the south of the Moray Firth in the Elgin area, and further south in the Midland Valley Middle ORS is entirely absent and Upper ORS rests on folded and faulted Lower ORS. A major period of faulting, folding and uplift of some basin areas clearly took place in Scotland during a phase

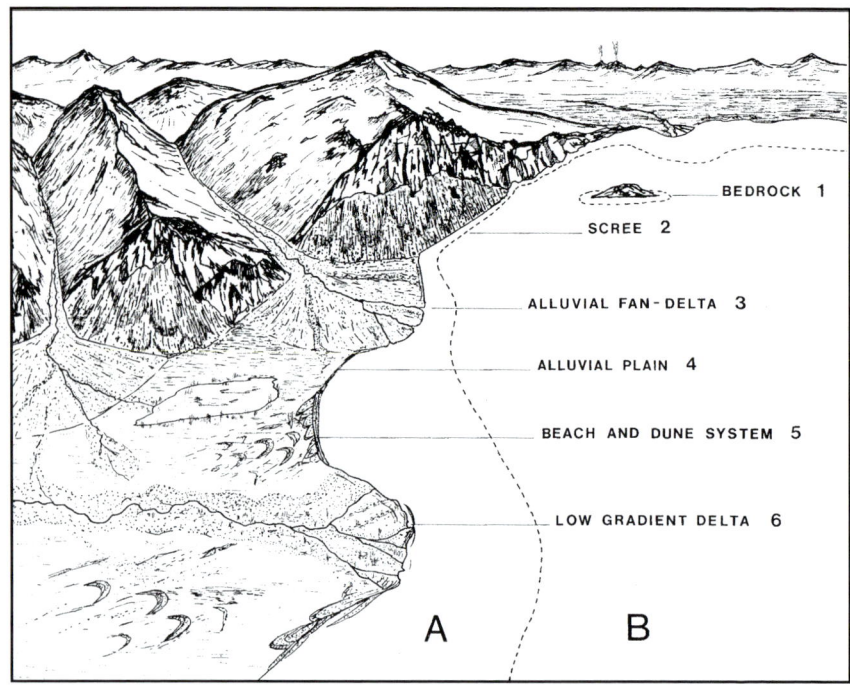

7 Diagrammatic sketch of marginal situations to the Orcadian lake at a time of high lake level. Zone A represents shallow areas in which bottom conditions were oxygenated and area B the deeper lake where anoxic conditions existed beneath a thermocline. Modified from Trewin (1986).

of compression towards the end of Middle Devonian times. In Shetland, the Walls and Sandness formations were subjected to intense folding and granitic intrusion at about this time (c.360–370 Ma) (Mykura, 1991). Thus, there is evidence of a major compressive episode in the evolution of the major strike-slip belt at this time.

In contrast, Rogers *et al.* (1989) contend that there is much room for doubt with regard to the traditional stratigraphy. On the basis of palynological evidence they discount the presence, in several areas, of an unconformity between Middle and Upper ORS. They regard extensional tectonics as the controlling mechanism for sedimentation, and have no need for periods of transpression. However, exposure is poor in the area, so making interpretation of the outcrop data ambiguous; a transpressive phase producing a gap in the section representing say two million years would be difficult to detect by palynology. Underhill and Brodie (1993) have interpreted seismic data from Easter Ross and favour early Devonian rifting followed by Middle to Late Devonian subsidence. There does not appear to be a marked break between Middle and Upper ORS in their seismic interpretation of this part of the basin. It is possible that in some basinal areas subsidence continued from Middle to Upper ORS, whilst transpressive uplift and erosion was more effective in basin marginal areas.

The Upper ORS consists dominantly of red–yellow sandstones of fluvial origin, and only rarely are fish remains found, apart from the well-documented faunal sequence of the Elgin–Nairn area summarised by Mykura (1991). The affinities of the fish faunas with Baltic forms reflects greater connection between river systems of northern Europe at this time, and a new basin configuration is inferred. The Upper ORS is present offshore in the Buchan Field where it passes conformably upwards into Lower Carboniferous rocks consisting of sandstones and shales of alluvial or deltaic aspect, which contain terrestrial plant debris. The fact that derived Carboniferous spores are found in Lias strata at Golspie (Windle, 1979) attests to the probable presence of Carboniferous in the Inner Moray Firth. Carboniferous strata have been described from the Outer Moray Firth by Leeder *et al.* (1990).

4. Post-Caledonian igneous activity
Devonian
Lavas and volcaniclastic rocks occur within the Middle ORS of Orkney and at the base of the Upper ORS, but no examples occur within the area of this guide.

Permo-Carboniferous
Within the area of the guide, there are a few examples of igneous activity which are generally assigned to the Permo-Carboniferous. Monchiquite dykes with a general ENE–WSW trend occur in the Dunnet Head region (see Fig. 1) and are roughly parallel to the trend of similar monchiquite and camptonite dykes in Orkney.

Two small vents are also present at Burn of Sinnigoe, Dunnet Head (Exc. 5, Loc. 15) and at Duncansby (Exc. 5, Loc. 7). They are filled with basalts, agglomerates and sedimentary clasts. The igneous rocks in the neck at Duncansby were described as nepheline basalts by Crampton and Carruthers (1914) and this vent has been dated as Permian (255 Ma by K/Ar method on basaltic material; see legend of BGS Sheet 116E, Wick). Although most clasts in the vents are of Old Red Sandstone lithologies, some clasts of basement gneiss have also been recorded.

Tertiary
No examples of Tertiary igneous activity are known in the area of this guide, but tuffs from volcanoes on the west coast contributed to Tertiary sediments in the North Sea, particularly in Palaeocene times.

5. The Inner Moray Firth Basin
The coastline from Inverness to Wick is strongly controlled by the Great Glen and Helmsdale faults which have played a major role in the development of the Inner Moray Firth Basin (Fig. 8). The presence of a dominantly Mesozoic basin had long been apparent from the marginal Mesozoic onshore outcrops, and presence of Mesozoic erratics transported onshore by ice during the Pleistocene glaciations. Work by the Geological Survey (summarised by Chesher and Lawson, 1983) identified the major structural elements, but most research was connected with the search for oil and gas, particularly following the discovery of the Beatrice Oilfield in 1976.

The Inner Moray Firth Basin is coincident with part of the Devonian Orcadian Basin and it is probable that the same basement faults influenced both basins. Following Devonian deposition, which probably continued into the Lower Carboniferous, a period of faulting, uplift and erosion occurred. Intense shearing affecting Devonian arkose and Helmsdale Granite along the Helmsdale Fault (Exc. 3, Itin. 4)

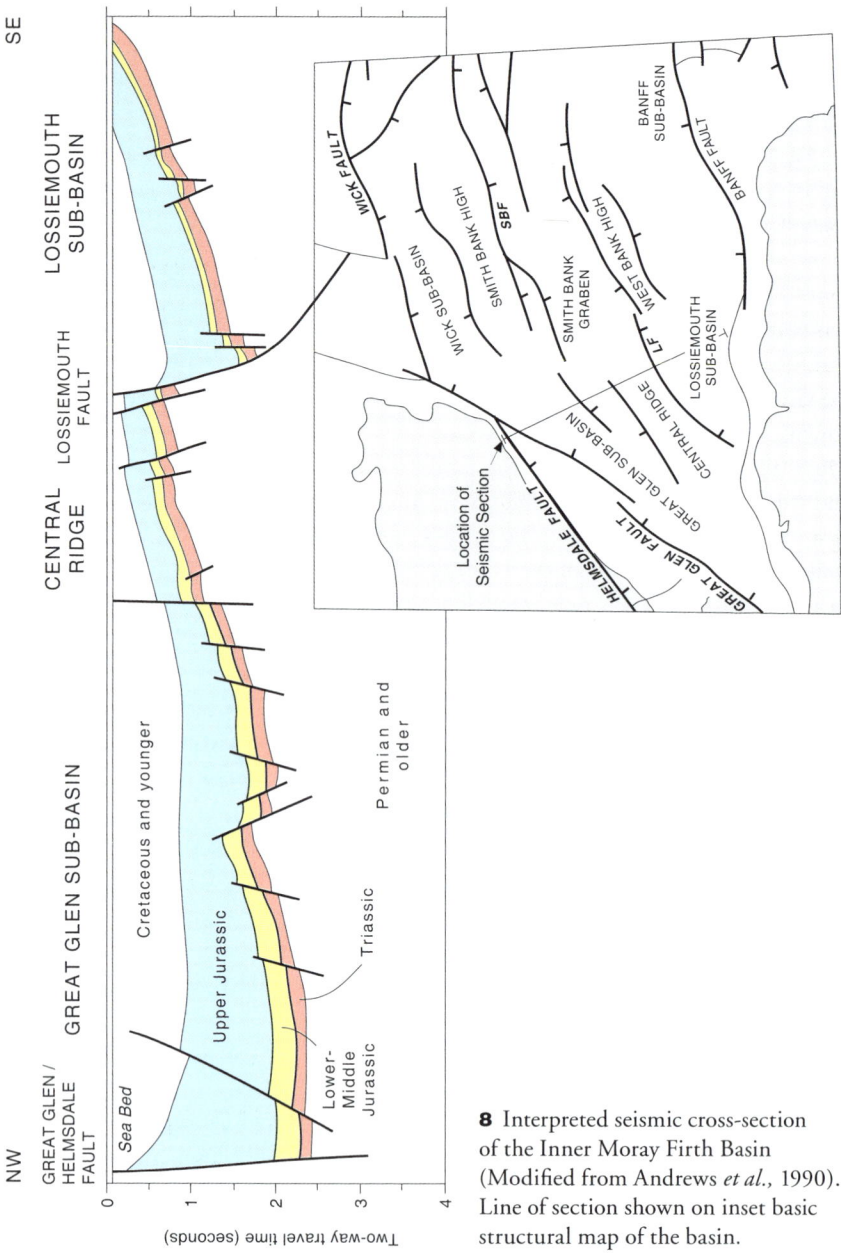

8 Interpreted seismic cross-section of the Inner Moray Firth Basin (Modified from Andrews *et al.*, 1990). Line of section shown on inset basic structural map of the basin.

is pre-Mesozoic in age and might relate to strike-slip displacement in Carboniferous times (Flinn, 1992). On the basis of seismic interpretation of the Easter Ross area Underhill and Brodie (1993) also recognise an event of compression and inversion in the Permo-Carboniferous.

Deposition resumed in Permian times with basin-marginal aeolian dune sandstones preserved at Hopeman to the south of the Moray Firth and offshore evidence of Permian evaporites in the basin centre (Taylor, 1990; Glennie, 2002). Triassic sediments are extensive in the Inner Moray Firth, and although Frostick *et al.*

(1988) proposed that the Great Glen Fault acted as a rift-margin fault controlling sedimentation, their view was not supported by evidence from the offshore area showing that the Great Glen Fault had no influence on thickness of the Permian to Jurassic section (Underhill, 1991).

The Wick and Banff fault systems define the north and south basin margins (Fig. 8) between which McQuillan *et al.* (1982) considered some 5 to 6 km of extension took place in Triassic to Cretaceous times, and was accommodated by dextral movement of the Great Glen and Helmsdale faults. Roberts *et al.* (1990) suggested that some 2 km of displacement may have formed a small Permian half-graben in the centre of the basin and that the later *c.* 6 km of displacement took place in the late Jurassic. Roberts *et al.* (1990) interpreted structure seen on seismic sections across the Great Glen Fault as indicative of strike-slip movement, but Underhill (1991) has convincingly demonstrated that the Great Glen Fault was effectively pinned during Mesozoic extension, and certainly had no effect on Jurassic deposition. Thomson and Underhill (1993) consider that rifting took place during the Permo-Triassic and again in the late Jurassic. The Helmsdale Fault was probably the main feature controlling sedimentation during both periods and formed the effective basin margin. In the Lower Cretaceous a shift in depocentres (Andrews *et al.*, 1990) reflects a change in the stress regime with different faults controlling deposition.

Underhill (1991) recognised that structures affecting Mesozoic rocks along the line of the Great Glen Fault are of Tertiary origin, and Thomson and Underhill (1993) related such tectonism to the geographical position of the Moray Firth between Atlantic rifting and Alpine fold-thrust tectonic regimes. Local transpression in the Tertiary resulted in the fold structures seen in the Upper Jurassic north of Helmsdale (Exc.3, Itin. 4.), which Thomson and Underhill (1993) relate to the interaction of sinistral strike-slip on the Helmsdale Fault and dextral movement on the Great Glen Fault. Further discussion of basin development is given in the following stratigraphic sections.

6. Permo-Trias

Deposits of the New Red Sandstone have a sporadic distribution onshore in northern Scotland, but are extensively represented offshore with thick sequences in the Minch and Hebrides basins against the Minch Fault, in the West Shetland Basin, and extensively within the Inner Moray Firth and the North Sea (Glennie, 2002).

Onshore deposits in the Moray Firth area are restricted to the Permian and Triassic of the Elgin area to the south of the Firth, and a small area of Triassic outcrop at Golspie (Exc. 1) in the area of this guide (Fig. 9). The present shoreline of the Moray Firth approximates to the margin of the Triassic and earlier Permian basin area, and coincides broadly with the earlier Devonian basin. Northern Scotland lay about 15° north of the equator in Permian times, increasing to more than 30° by late Triassic. The climate was semi-arid to arid and deposition was dominated by flood deposits of fluvial systems near the land areas (e.g. Burghead Beds), together with aeolian sandstones (e.g. Hopeman Sandstone, Lossiemouth Sandstone). Basin centres received the finer detritus of sand and mud with resulting fine sandstones and mudstones deposited on widespread alluvial plains and in playa-lakes.

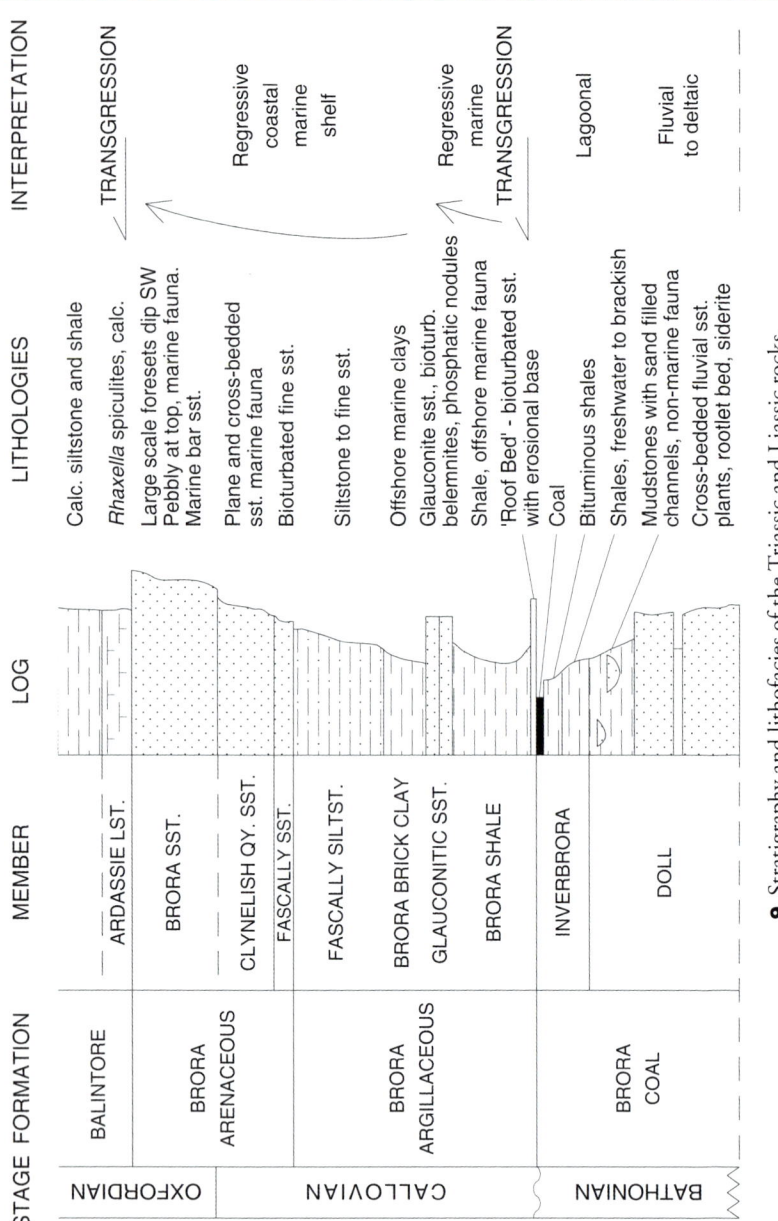

9 Stratigraphy and lithofacies of the Triassic and Liassic rocks.

The dominantly aeolian Hopeman Sandstone (Glennie, 2002) forms the lower part of the Permo-Triassic succession to the south of the Firth, and the discovery of a *Dicynodon* skull (Clark, 1999) dates this sandstone to the Late Permian. Coastal outcrops of the Hopeman Sandstone consist of upper and lower dune-bedded units separated by a unit of distorted and disrupted dune bedding with evidence of reworking by water. Glennie and Buller (1983) have suggested that these disruption structures were formed during the Zechstein (Upper Permian) marine transgression. Marine Zechstein occurs offshore in the Moray Firth (e.g. wells 12/23-1 12/30-1; Taylor, 1990) and a dune belt could have bordered this arm of the Zechstein sea.

The rest of the Permo-Triassic succession of the Elgin area comprises the dominantly aeolian Lossiemouth Sandstone, which has yielded Upper Triassic reptiles (Benton and Walker, 1985), and the fluvial Burghead Beds. These Triassic sandstones and pebbly sandstones are about 150 m thick in the Elgin area, but thicken to 500 m, and are much finer-grained in the offshore area, containing siltstones, mudstones, thin limestone and evaporite beds (Frostick et al., 1988). Deposition in the basin centre may have taken place in ephemeral lakes.

To the south of the Moray Firth the top of the Trias is marked by the Stotfield Cherty Rock. This rock is believed to be a fossil soil profile formed in semi-arid conditions and is probably an altered calcrete rather than a silcrete (Frostick et al., 1988). The horizon is widespread in the Inner Moray Firth and is present as a calcrete at Golspie, where it overlies Triassic mudstones and sandstones (Exc. 1). The Triassic sandstones at Golspie are mainly waterlain but contain abundant well-rounded quartz grains formed by aeolian transport, and some cross-bedding is of probable aeolian origin. These rocks are similar to the Lossiemouth Sandstone to the south of the Moray Firth. The fact that the calcrete forms an extensive seismic marker horizon in the Moray Firth (Linsley et al., 1980; Underhill, 1991) indicates that at the end of the Triassic deposition was slow, the climate semi-arid, and surrounding topography subdued. The Great Glen Fault certainly did not form an active rift margin at this time, and deposition was continuous across the line of the fault.

Thus, at the close of the Triassic, the relatively subdued Grampian and Highland areas bordered an extensive spread of alluvial plain and playa deposits in the Moray Firth. As Europe moved further away from the equator a major climatic change took place to a more temperate climate, and the Triassic basins were invaded by the shallow shelf seas of Jurassic times.

For further information on the Elgin area the reader can consult Glennie (2002) in 'The Geology of Scotland', the Elgin memoir (Peacock et al., 1968), Glennie and Buller (1983), Frostick et al. (1988) and the excursion guide by Gillen (1987).

7. Jurassic

Introduction

The coastal outcrop of Jurassic rocks from Golspie to north of Helmsdale at Dun Glas is bounded to the NW by the Helmsdale Fault (Fig. 1). The local base of Jurassic is generally considered to lie below the Dunrobin Pier Conglomerate, but the contact with the underlying Triassic calcrete is not exposed. It is possible that Rhaetic strata are present, since the marine transgression from the south, and a change to a wetter climate, commenced in the late Triassic. The general succession (Fig. 2), which ranges from Lower Jurassic (Hettangian) at Golspie in the south to Upper Jurassic (Volgian) to the north of Helmsdale, is virtually complete apart from a gap in the Lower–Middle Jurassic where Toarcian and Bajocian strata are not represented. The gap could be due to non-exposure or strata cut out by faulting, but in view of the known offshore stratigraphy it is likely to be due to unconformity caused by uplift of the North Sea Dome and associated Jurassic volcanic activity. (Underhill and Partington, 1993; Underhill, 1998). Another possible gap is present in the Upper Oxfordian, but this cannot be tested due to lack of exposure. The

Jurassic strata are covered in Excursions 1, 2 and 3 to which the reader is referred for geological detail. The Jurassic history of the area falls into two sections. The first concerns all the Jurassic strata up to the Ardassie Limestone of the Balintore Formation (Mid–Late Oxfordian) which constitute a great variety of lithologies representing environments ranging from alluvial to lagoonal and shallow marine. The second section is the dominantly Kimmeridgian succession of shales, sandstones and boulder beds which were deposited on the downthrow side of the active Helmsdale Fault, and form some of the most spectacular deposits of the British Jurassic. For a general account of the Jurassic of Scotland see Hudson and Trewin (2002).

Whilst the Jurassic of Brora–Helmsdale is the main onshore outcrop area, there are small exposures of parts of the succession against the Great Glen Fault at Eathie (Waterston, 1951; Wignall and Pickering, 1993), near Balintore and Port an Righ (Sykes, 1975a), and to the south of the Moray Firth in the Lossiemouth borehole (Berridge and Ivimey-Cook, 1967). The Inner Moray Firth is a Mesozoic basin area and remains an oil exploration area, although the only major commercial success to date is the Beatrice Oilfield discovered by Mesa Petroleum in 1976 (see Economic Geology). Offshore drilling and seismic surveys have provided much information on the offshore Jurassic, which has been summarised by Glennie (1998). Useful contributions to the understanding of the Moray Firth area have been by Andrews and Brown (1987), Andrews *et al.* (1990), Underhill (1991) and Stephen *et al.* (1993).

Hettangian to Mid-Oxfordian

The Lias stratigraphy of the Golspie area (Fig. 9) was revised by Batten *et al.* (1986) who obtained a flora from the Dunrobin Pier Conglomerate (Exc. 1) which had previously been considered unfossiliferous (Neves and Selley, 1975). The stratigraphic nomenclature has been subsequently modified by Richards *et al.* (1993) (see also Excursion 1). This conglomerate contains numerous clasts from the underlying Triassic calcretes and sandstones which had presumably been locally uplifted and eroded, possibly by early movements on the Helmsdale Fault. The conglomerate was deposited in an alluvial fan environment with current transport to the NE; it does not occur in offshore boreholes, and is probably restricted to the basin margin. The flora present is of terrestrial origin and is probably of Hettangian age, but a Rhaetian age is not entirely excluded. This deposit marks a change to a more humid climate with evidence of a rich vegetation cover on the Scottish landmass.

The succeeding strata of the Carbonaceous Siltstone and Clay Unit of the Golspie Formation are very seldom exposed, but from borehole evidence deposition under dominantly freshwater conditions in an alluvial plain environment has been suggested (Neves and Selley, 1975). Within this succession influxes of dinoflagellates indicate minor marine incursions, possibly associated with brief establishment of lagoonal conditions. Similar rocks rest on the Triassic offshore in the Beatrice Oilfield and also occur at Lossiemouth to the south of the Firth.

The White Sandstone Unit (= Mains Formation, Fig. 9) is a medium- to coarse-grained cross-bedded sandstone which separates dominantly freshwater deposits from overlying marine shales and thin sandstones of the Lady's Walk Shale Member. Deposition of the White Sandstone took place at, or close to, the contemporary coastline; it is interpreted as the deposits of an estuarine channel environment. Land-derived plant debris is abundant together with marine bivalves (Lee, 1925) and marine microplankton (Neves and Selley, 1975).

Above the White Sandstone, the Lady's Walk Shale Member is of shallow marine origin, and records the first Jurassic marine invasion of the Inner Moray Firth in Sinemurian times. The shales range up to Pliensbachian age on the basis of a sparse ammonite fauna. A great variety of lithologies is present, but the sandstones contain the richest marine fauna, which includes bivalves (oysters and pectinids) and rhynchonellid brachiopods. Individual beds have distinctive faunas which were suited to prevailing bottom conditions. In silty shales and mudstones at the top of the exposed section shallow sand-filled scours containing quartz pebbles, wood fragments, and marine bioclastic debris are present, and attest to rapidly changing environments and proximity to land. The numbered bed sequence given by Lee (1925) generally cannot be followed due to poor exposure caused by a cover of beach sand. In a regional context, the marine invasion of the Inner Moray Firth is later than that seen on the west coast of Scotland, where open marine conditions were established in Hettangian times in the Hebrides basins (summary in Hudson and Trewin, 2002).

Toarcian and Bajocian age sediments are not exposed onshore. Offshore the sandstones of the Orrin Formation, although assigned to the Toarcian–Bajocian by Andrews and Brown (1987) are now considered to be of Pliensbachian to Toarcian age (Stephen *et al.*, 1993). These sandstones show an upward transition from marine shoreface enviroments to those of a beach barrier and interdistributary bay complex (Stephen *et al.*, 1993).

The Bathonian Doll Member of the succeeding Brora Coal Formation (Fig. 10 and Exc. 2) is of fluvial origin with channel sandstones and alluvial plain mudstones (Hurst, 1981). Siderite and abundant kaolinite are present as well as a sparse fauna of freshwater ostracods; the freshwater bivalve *Unio* was recorded from the top of the Member by Neves and Selley (1975). Drifted plants occur in the shales, some of which, including several species of *Equisetum* (mare's-tails) together with *Ginkgo*, *Goniopteris*, *Todites* and *Cladophlebis*, formed the subject of Marie Stopes' first palaeobotanical paper (Stopes, 1907). Offshore in the Beatrice Oilfield similar kaolinitic and siderite mudstones with thin sandstones containing rootlet casts are present.

The Inverbrora Member of the Brora Coal Formation is considered to be of lagoonal origin. Lam and Porter (1977) first noted the presence of marine microplankton elements (dinoflagellates) and a study by MacLennan and Trewin (1989) showed that marine influence is extensive with high abundance/low diversity dinocyst assemblages present. Shell beds of *Neomiodon* and *Isognomon* together with ostracods were winnowed on the lagoon floor during phases of marine invasion. The sediments are dominated by dark organic-rich shales (up to 27% Total Organic Carbon) which locally approach oil-shale in character and were deposited during stagnant lagoonal phases. Pyrite is abundant, probably reflecting that abundant sulphate was periodically available from sea water.

In the Beatrice Oilfield, this lagoonal facies is absent or very thin, but it is represented to the south of Brora at Cadh' an Righ, with similar lithologies, fauna and palynological character to those of Brora (MacLennan and Trewin, 1989). The

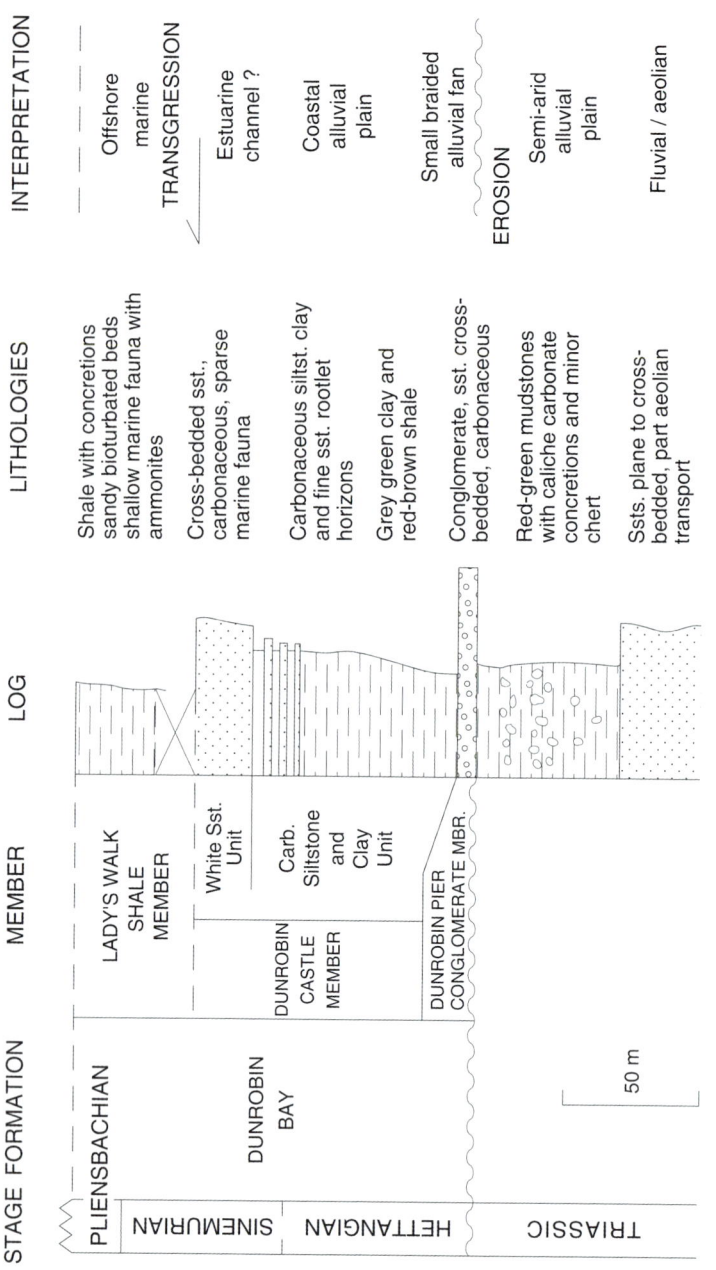

10 Stratigraphy and lithofacies of the Bathonian to Oxfordian section.

lagoonal area (Fig. 11) probably extended parallel to the Helmsdale Fault system and was periodically invaded by the sea from the NE in the region of the Wick Fault, with the ocean connection through the Viking Graben. An alternative explanation is that the sea entered the Moray Firth through the Great Glen from the west coast.

The Brora Coal overlies the lagoonal Inverbrora Member. This coal, which has been extensively worked in the Brora area (see Economic Geology Section) is also present in the Beatrice Oilfield and onshore to the south at Cadh' an Righ. The coal generally lacks a seat earth, initial deposition being from drifted plant material. The

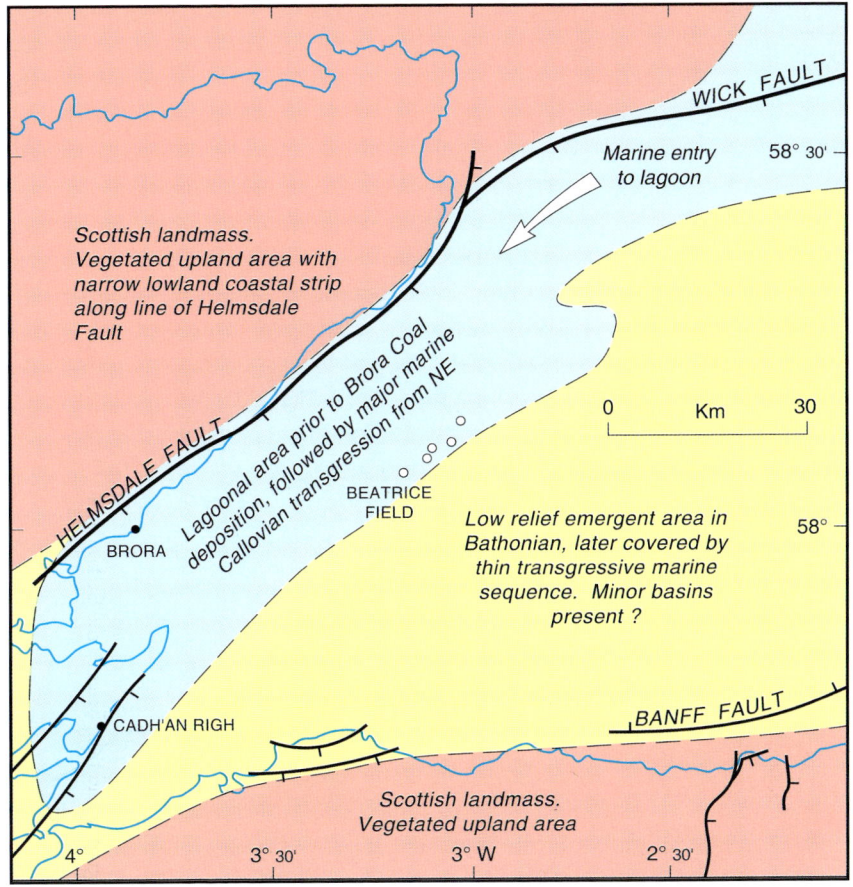

11 Generalised palaeogeography of the Inner Moray Firth during the early Callovian, showing the probable extent of the lagoonal area (After MacLennan and Trewin 1989).

lagoonal area was cut off from the sea at this time and swamp conditions, probably similar to a floating bog, spread over the lagoon area. The coal contains coniferous wood and masses of *Equisetum* (Harris and Rest, 1966). Palynology preparations of the coal are highly variable, representing local plant communities, and marine microflora is absent (MacLennan and Trewin, 1989).

The Bathonian/Callovian boundary has frequently been positioned above the coal, but from palynological data, probably lies within the Inverbrora Member (MacLennan and Trewin, 1989). Overlying the coal at Brora, at the base of the Brora Argillaceous Formation (Fig. 10), is the Brora Roof Bed, a bioturbated shallow marine transgressive sandstone with a shelly fauna dominated by bivalves. This bed marks the main Callovian marine transgression in the area (Sykes, 1975a). Reworked coal fragments occur in the Roof Bed, and to the south of the area at Cadh' an Righ sand-filled burrows extend down into the coal from the Roof Bed. Rapid deepening of the sea resulted in deposition of the Brora Shale with a diverse marine macrofauna of bivalves together with belemnites and ammonites. The environment was clearly open marine at this stage, but a high proportion of land-derived spores and plant debris attests to the close proximity of the Scottish landmass.

In the Beatrice Oilfield the coal is followed by a few metres of shale in which the transition to open marine conditions takes place. Several (usually four) thin sandstones, generally forming coarsening-up sequences representative of marine bars, appear to be the broad time equivalent of the Roof Bed (the 'B' sandstones of the Beatrice Reservoir, Fig. 12). The Mid-Shale overlies the 'B' sandstones and compares well with the Brora Shale, with similar faunas and palynofacies characteristics (MacLennan and Trewin, 1989). A general comparison of this part of the Brora section with that of Cadh' an Righ and Beatrice is given in Figure 12.

Overlying the Brora Shale at Brora is the Glauconitic Sandstone Member, which contains highly glauconitic beds also rich in siderite and phosphatic concretions. Belemnites are particularly abundant and the sandstones are extensively bioturbated. This member represents a period of slow deposition at Brora, but to the south at Cadh' an Righ a nodule bed represents a condensed sequence, or possibly a non-sequence. In Beatrice Oilfield wells, phosphatic concretions present near the base of the 'A' sandstone are probably the lateral equivalent of the Glauconitic Sandstone Member.

At Brora the rest of the Brora Argillaceous Formation (Fig. 10), together with the Brora Arenaceous Formation, forms a marine coarsening-up sequence commencing with offshore marine clays of the Brora Brick Clay Member and culminating in the cross-bedded porous, pebbly sandstones with moulds of marine bivalves typical of the Brora Sandstone. These sandstones were probably deposited in a coastal marine bar system oriented parallel to the Scottish landmass. A similar coarsening-up marine bar sequence occurs in the Beatrice Oilfield where it forms the main oil reservoir, the 'A' sandstones (Fig. 12). However, the coarsening-up sequences of Brora and Beatrice are not synchronous, the top of the Brora sequence being Oxfordian (*plicatilis* Zone) and that at Beatrice Callovian (*lamberti* Zone) (Fig. 11). Contrasting subsidence rates and sand supply in the two areas may account for the difference.

This part of the Jurassic sequence at Brora is completed by the transgressive Ardassie Limestone Member (Fig. 9) of the Balintore Formation, which consists of shales and muddy limestones with calcitised *Rhaxella* (sponge) spicules. These rocks are intensely bioturbated with *Chondrites*, *Thalassinoides* and other trace fossils; they also contain thick-shelled oysters (*Gryphaea dilatata*), fan mussels (*Pinna lanceolata*) and belemnites. Ammonites of *plicatilis* Zone were recorded by Sykes (1975a). Offshore in the Beatrice Oilfield the Oxfordian is represented by marine shales, but further offshore the Alness Spiculite Formation is present and resembles the Balintore Formation with its concentrations of *Rhaxella* spicules (Andrews and Brown, 1987; Stephen *et al.*, 1993).

When the sequence from the base of the Lias to the Balintore Formation is traced across the Inner Moray Firth a gradual thinning is observed from the Brora area eastwards across the Great Glen Sub-basin to the Central Ridge (Andrews and Brown, 1987; Underhill, 1991). This thinning continues across the Lossiemouth Fault and the Lossiemouth Sub-basin to the Moray coast (Fig. 8). Thus, in this part of the Jurassic only the Great Glen Sub-basin was present, and maximum thickness occurs close to the Helmsdale Fault on the NW of the basin. There is no evidence that the Great Glen Fault had any significant effect in controlling sedimentation; from seismic data (Underhill, 1991) the sequence thickens into the Helmsdale Fault.

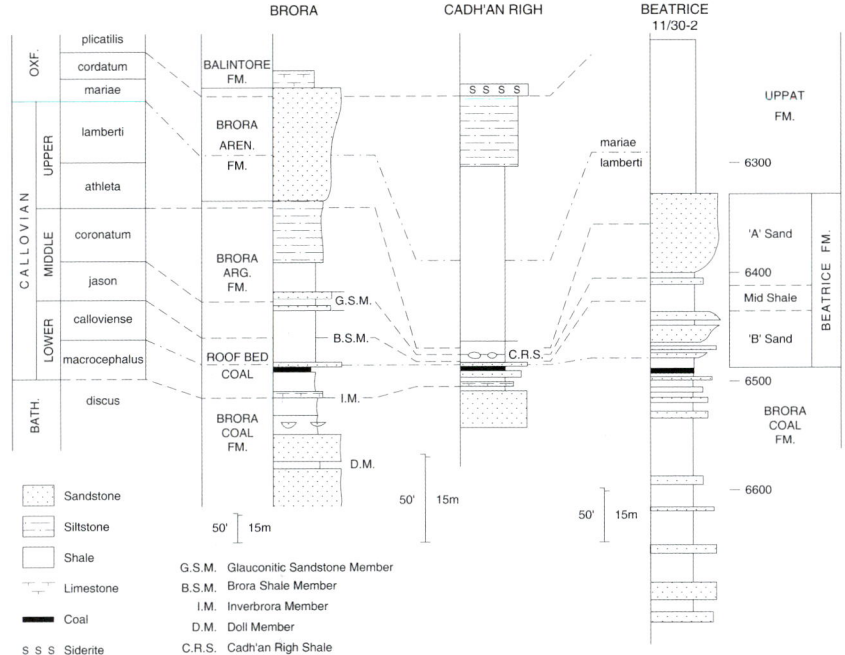

12 Stratigraphic sections, nomenclature and correlations for the Callovian at three Inner Moray Firth Basin localities. Note variation in scales (Modified from MacLennan and Trewin 1989).

Late Oxfordian to Ryazanian

In the outcrop area covered by the guide this part of the succession comprises shales, siltstones, sandstones and boulder beds dominantly of Kimmeridgian age. Active synsedimentary faulting on the Helmsdale Fault and other faults in the Inner Moray Firth (Smith Bank Fault, Lossiemouth Fault) resulted in deposition of up to 3,000 m of strata which are thickest close to the downthrow side of the active faults (Fig. 8). From Kintradwell to the north of Helmsdale the effects of synsedimentary fault movement on sedimentation can be examined in great detail, and are covered in the itineraries of Excursion 3.

The succession (Fig. 2 and Exc. 3, Fig. 3.2) is at least 800 m thick. There is a general younging of strata northwards along the coast. Stratigraphic work on the ammonites (Bailey and Weir, 1932; Linsley, 1972; Brookfield, 1976; Wignall and Pickering, 1993) and palynology (Lam and Porter, 1977; Riley, 1980; Barron, 1989) indicates that the succession is probably continuous from *cymodoce* Zone of the Kimmeridgian to *albani* Zone of the Middle Volgian. Whilst there is some evidence of early Jurassic activity on the Helmsdale Fault (see above), the fault does not seem to have exerted a strong control on sedimentation until Kimmeridgian times, when the fault scarp formed a submarine topographic feature.

At Brora, little is known of the Upper Oxfordian rocks that overlie the Ardassie limestones due to lack of exposure. Borings at Ardassie Point indicate that bio-turbated shales with thin sandstones deposited in a shallow marine environment are present. The first exposures to the north of Brora at Kintradwell are shales of *cymodoce* Zone which have interbedded boulder beds and slumps derived from a shallow shelf to the west. At some time in the late Oxfordian or early Kimmeridgian the fault became active, downthrow to the east was rapid, and sedimentation could not keep pace with subsidence. The result was the rapid creation of an exposed submarine fault scarp with a deepwater marine environment on the downthrow side of the fault, and a narrow shallow-marine shelf and shoreline on the upthrown side (Exc. 3, Figs. 3 and 17). This shelf is not preserved but its presence can be deduced from the evidence of fauna, flora and clasts that were swept off the shelf into deep water to be preserved along with the open marine fauna of ammonites and belemnites (see Exc. 3).

In the region of Allt na Cuile and Lothbeg Point boulder beds are replaced by a porous quartzose sandstone interbedded with fissile siltstones and shales. The siltstones and shales contain ammonites of *cymodoce* Zone and are in part the lateral equivalents of the boulder beds of Kintradwell (Wignall and Pickering, 1993). Well-preserved leaves of land plants are found in association with the ammonites in this area, and provide evidence of the flora of the adjacent landmass. Van der Burgh and van Konijnenburg-Van Cittert (1984) recognised plants typical of brackish swamps (*Gleichenites cycadina*) and freshwater swamps (*Taxodiophyllum scoticum*) of a low-lying delta to be the dominant floral elements. Less abundant representatives of heath, moist lush vegetation and upland forest were also recognised. These sandstones (Allt an Cuile Sandstone) were derived from a delta area on the upthrown side of the fault which developed at a river mouth draining the Scottish landmass. The sandy delta deposits spilled over the fault line to form sandy submarine deposits in the deep water. Evidence of channeling, downslope sand movement and deposition from turbidity currents is seen. Vegetation swept out by floods became waterlogged and sank into deep water close to shore to be preserved along with the marine fauna. The water in which the sandy fan accumulated was deeper than local wave base, since wave-produced structures are absent, but bioturbation present in underlying shales and in some sandstones implies moderately shallow conditions. The Lothbeg Siltstone overlies the Allt na Cuile Sandstone, and is of *mutabilis* Zone age (Wignall and Pickering, 1993). Drifted land plants are common in these rocks and benthic faunas record an upward decrease in benthic oxygenation levels which is continued in the overlying Helmsdale Boulder Beds (Wignall and Pickering, 1993).

The rest of the Jurassic coastal sequence to the north of Helmsdale (Exc. 3) displays the interbedding of boulder beds, dark siltstones and shales which comprise the Helmsdale Boulder Beds. Close to the Helmsdale Fault chaotic deposits of rock-fall breccias are seen (e.g. Exc. 3, Itin. 3) forming a strip up to a few hundred metres in width. Further from the fault, boulder beds are interbedded with dark fissile siltstones and thin pale sandstones. In some places giant boulders are seen, the most spectacular being the (wrongly named) 'fallen stack' at Portgower (Exc. 3, Itin. 3). This clast of bedded Middle ORS flagstones is at least 30 m long, and must have fallen and slid into deep water from an exposed fault scarp greater than 30 m in height. Most boulder beds were deposited as matrix-poor debris flows and some can be observed to wedge out away from the fault. The blocks in the boulder beds frequently exceed the bed thickness and protrude above the bed top; lamination beneath the beds shows distortion caused by emplacement of the beds.

Thin sandstone beds within the dark siltstones contain plant debris and marine bioclasts which were probably deposited when storms battered the coastline and swept sand over the fault scarp from the shallow shelf (Exc. 3, Fig. 3). In the Helmsdale area the boulder beds have a matrix of bioclastic debris including fragments of bivalves, echinoids, brachiopods, serpulid worms and other shallow marine organisms. Coral colonies (*Isastraea*) are also present, and some clasts were bored by bivalves prior to incorporation in the boulder beds. In this area a rocky shelf with high organic productivity bordered the fault scarp, but little detrital sand was available, in contrast to conditions to the south at Kintradwell where boulder beds have a sandy matrix.

Movements on the fault must have resulted in severe earthquakes which probably triggered the slumps that are seen in both the Kintradwell and Helmsdale boulder beds. Sandstone dykes (Exc. 3, Itin. 1) were formed when sand, liquefied by shock, was injected into fractures. Pickering (1983, 1984) has described some of the sedimentological detail of these deposits.

Along the line of the fault there are distinct changes in clast types in the Boulder Beds. These were discussed by Bailey and Weir (1932) and have been investigated by MacDonald (1985, and Exc. 3 in this guide). In the south of the area some reworked Jurassic material (sand and pebbles) is present together with clasts of cross-bedded sandstones of fluvial origin, which, on the basis of petrographic comparisons, are probably derived from the upper ORS. From Portgower to Helmsdale clasts of Middle ORS flagstones dominate. On the upthrow side (footwall) of the fault in the early Kimmeridgian, it is possible that poorly consolidated Jurassic strata rested directly on Upper ORS fluvial sandstones which overlay Middle ORS flagstones. At present, Moine psammites, Lower ORS and Helmsdale Granite occur to the west of the fault. Clearly these rocks were not unroofed, nor were they exposed on the Helmsdale Fault scarp in the Kimmeridgian, since no clasts of these rocks occur in the boulder beds. Thus considerable post-Kimmeridgian movement has taken place on the Helmsdale Fault.

Whilst there may still be minor differences in interpretation, the general story of fault-controlled sedimentation of the boulder beds follows the classic interpretation of Bailey and Weir (1932) which has been endorsed by more recent work (Crowell, 1961; Linsley, 1972; Pickering, 1984; MacDonald, 1985; Wignall and Pickering, 1993; Underhill, 1994). Earlier interpretations of the origin of the boulder beds include crush breccias (Murchison, 1827), penecontemporaneous coastal erosion (Cunningham, 1841) ice transport (Ramsay, 1865), violent floods from rivers (Judd, 1873), screes with an ice foot (Blake, 1902), and falls from steep hillsides into the sea (Woodward, 1911). Hugh Miller came very close to the current opinion, recognising the Devonian fish in the clasts, which he considered had been derived from a nearby mountainous area (Miller, 1854) and he also recognised the role of earthquakes in the formation of the sandstone dykes (Miller, 1859). The 'fallen stack' at Portgower was first interpreted as such by Blake (1902) and this was repeated by Macgregor (1916), at which time steep cliffs or hillsides were envisaged as the source of boulders. Bailey and Weir visited the area in 1930 and 1931 following earlier realisation by Bailey of the true origin of the boulder beds. Their careful fieldwork and masterly exposition ended major speculation on the origin of these beds. Bailey and Weir acknowledge the help with fieldwork of two students of the Glasgow honours class – A. Lamont and J.G.C. Anderson.

It is unfortunate that the boulder beds cannot be followed for a greater distance away from the fault. Twenty kilometres offshore at Beatrice the succession is dominated by shale with only thin sand beds; it is probable that the debris flow beds only extend for a few kilometres offshore.

8. Cretaceous

No undoubted Cretaceous strata occur onshore in situ in the area covered by this guide, but Lower Cretaceous crops out on the seabed over large areas of the Inner Moray Firth. A summary of the Cretaceous history of Scotland is provided by Harker (2002). Active faulting in the Inner Moray Firth was a major control on deposition and it is probable that the Helmsdale Fault was still active. The Lower Cretaceous is over 750 m thick within 10 km of Wick, but the axis of sedimentation has now moved away from the line of the bounding Helmsdale–Wick fault system. The thickness distribution of Lower Cretaceous does not mirror that of the fault-controlled Upper Jurassic and it is clear from isopach maps that different faults were active (Andrews *et al.*, 1990, Figs. 31, 34). An unconformity is generally recognised beneath the Lower Cretaceous.

The deposits are of fine to coarse sandstones and dark shales and mudstones. Sandstones deposited on the downthrow (hanging wall) sides of faults are generally interpreted to be of submarine fan origin, but sandstones with a marine shelly fauna were probably typical of shallow areas, particularly around the basin margins. It is probable that northern Scotland was still emergent and acting as a source of sand supply to the Inner Moray Firth.

The Upper Cretaceous Chalk Group is at seabed on the downthrow side of the Wick Fault only 20 km offshore (Andrews *et al.*, 1990). The Cenomanian transgression resulted in marine inundation and it is probable that, at times of highest sea level, northern Scotland was totally submerged or at least reduced to small islands. Marginal glauconitic sandstone 'Greensand' deposits occur in the Inner Moray Firth and are overlain by chalk and marl lithologies. Minor unconformities and subsequent onlap within the Upper Cretaceous of the Moray Firth reflect changes in sea level. Thickness of the Upper Cretaceous is controlled by the position of pre-existent basins, and by continued movement on the Wick and Banff fault systems.

Onshore evidence of Cretaceous strata is limited to glacial erratic material scraped off the floor of the Moray Firth and deposited onshore by ice moving NW over Caithness and SE over Buchan. Thus, a large Lower Cretaceous sandstone erratic (Tait, 1912) at Leavad in Caithness which contained a shelly fauna in calcareous concretions formed the basis for a sand quarry. It is not entirely clear whether the erratic was a single intact block or a composite broken mass. Numerous smaller Cretaceous erratics, particularly concretions from Lower Cretaceous sandstones and flints from the Upper Cretaceous, have been recorded in Caithness.

The only *in situ* occurrence that might be of Cretaceous age is a slice of fossiliferous sandstone within the Helmsdale Fault zone north of Helmsdale (Exc. 3, Itin. 4).

9. Tertiary

No Tertiary deposits are preserved in the area, but it is probable that early Tertiary fluvial and deltaic deposits were present in the Inner Moray Firth. Sediments were derived from the Highlands by rivers draining the tilted late Cretaceous marine planation surface which was created by Upper Cretaceous marine transgression. Opening of the N. Atlantic, and volcanic activity on the west coast (summary in Bell and Williamson, 2002) resulted in post-rift basin collapse in the North Sea and tilting of the old erosion surface due to rifting on the west coast of Scotland. The evidence for the previous existence of an early Tertiary delta in the Inner Moray Firth is preserved as delta-derived slope facies and deep-water sandstones in the Outer Moray Firth and Central Graben (summary in Knox, 2002). Uplift in the Inner Moray Firth, for which evidence is seen along the Great Glen Fault (Underhill, 1991; Thomson and Underhill, 1993), resulted in reworking of the fluvial to deltaic system and a shift of deltaic facies to the Outer Moray Firth. Data presented by Hurst (1982), and Pearson and Watkins (1983) using clay minerals to elucidate burial history indicate that up to 1 km of section was probably removed by Tertiary erosion in the Inner Moray Firth; an unknown proportion of this section was probably early Tertiary in age. The land areas over which Tertiary sediments were derived became deeply weathered, particularly in the sub-tropical climate of the early Tertiary, and again in the Miocene. Relics of Tertiary deep weathering are common in the Buchan area to the south of the Moray Firth (Kneller, 1987) but most have been stripped off by glacial erosion in the area of this guide, although evidence of deep weathering is still preserved in the Helmsdale area (Boulton *et al.*, 2002). Within Tertiary time there were several phases of regional tilting and many transgressive/regressive episodes. Remnants of tilted planar erosion surfaces in northern Scotland are ascribed to such events. The surfaces are attributed to marine planation by some authors whilst others consider them to be of subaerial origin.

10. Quaternary

It is difficult to unravel the record of growth and decay of the Quaternary ice sheets that were responsible for moulding much of the scenery of the Highlands. An excellent overview covering the whole of Scotland is given by Boulton *et al.* (2002), and the earlier work of Charlesworth (1956) and Sissons (1968) is also of interest. For specific field evidence relating to Caithness, and a historical summary, the observations of Omand (1973) are most useful. The difficulty with interpretation on land is due to the fact that the last glaciation, in the late Devensian and culminating about 20,000 years BP, removed or masked much evidence of earlier glacial periods.

In Caithness there is evidence of earlier glacial episodes even if the absolute ages cannot be determined with certainty. Omand (1973) recognised an early till (Dunbeath Till) of local derivation and lacking shells. This is overlain by a shelly till (Lybster Till) which contains molluscan shells of marine origin, and locally abundant Mesozoic glacial erratics derived from the bed of the Moray Firth. The most celebrated erratic is the Lower Cretaceous sandstone erratic at Leavad (Tait, 1912) which was apparently 220 x 137 m in area and up to 8 m thick. Other erratics include Jurassic lithologies, similar to those at outcrop, and flints of Upper Cretaceous origin.

In Buchan, to the south of the Moray Firth, similar Mesozoic erratics and shelly boulder clay were also derived from the Moray Firth. Thus, it is postulated that an ice dome existed in the Moray Firth at this time with ice flowing away from the centre of the dome, to the NW over Caithness and SE over the Buchan area. Further evidence for NW ice flow in Caithness is provided by a train of erratics of Sarclet Conglomerate extending from Sarclet NW to the Thurso and Dunnet areas on the north coast. This ice movement phase is thought to be older than 40,000 years on the basis of carbon dating (limit of method) of shell material, and could possibly be as old as a cold phase about 70,000 years ago.

The westward limit of the shelly till follows a line from Berriedale to Reay. To the west of this line, an upper till (Reay Till) of local derivation is found which may have been contemporaneous with the shelly till. The ice transporting this till appears to have flowed to the E and NE and been deflected northwards by the ice responsible for deposition of the shelly till.

The last glacial episode in Caithness in the Late Devensian, about 20,000 years ago, is not associated with extensive till deposits, and it is possible that Caithness lay at the margin of the ice sheet that covered most of Scotland. The Moray Firth ice dome was not re-established at this time and ice flowed freely into the Moray Firth. Moraines and fluvio-glacial gravels that overlie the local (Reay) and shelly till in Caithness represent marginal deposits of the ice sheet. The major valleys such as Langwell and Berriedale contain moraines of local material probably left from small valley glaciers which may have been fed from small, local ice accumulations. For example, glaciers in valleys at Dunbeath, Berriedale and Langwell may have been fed from a snowfield on Knockfin.

Periglacial processes in Caithness were severe, often disturbing the top metre of drift sections by frost action. The ice-smoothed landscape was further affected by solifluction, further filling and smoothing the valleys. The general undulating topography of Caithness with its cover of impermeable boulder clay was ideal for establishment of extensive post-glacial lakes and subsequent peatbogs.

Although there are no well-marked shorelines in the north of the area, relics of raised platforms can be identified south of Helmsdale, and impressive river terraces are present, particularly on the Brora River. The fine cliffs of the east Caithness coast have a history dating back beyond the Late Devensian glacial maximum. It seems probable that they were created by the outward flow of ice from the earlier Moray Firth ice domes. The ice easily eroded the soft Mesozoic strata, but was less effective at removing the Helmsdale Granite and Devonian strata on the NE side of the Helmsdale Fault. Perhaps the ice was initially deflected NE up the coast prior to over-riding Caithness as the ice accumulation developed. The cliffs are preserved because the land-derived Late Devensian ice sheet did not cover them, only sending minor glaciers down the major valleys which cut the coastal cliff line.

11. Economic geology

This area is remarkable for the variety of geological features that have been, or still are being exploited. Interesting historical information can be found in Read *et al.* (1925) and Crampton and Carruthers (1914), whilst more recent information is given in Johnstone and Mykura (1989) and Beveridge *et al.* (1991). Much of the following information is based on these accounts.

The flagstone industry

The effects of the exploitation of the flagstones of the Middle Old Red Sandstone of Caithness have left a unique legacy to the whole county. The upright slab walls of field divides, large roof flags on croft outhouses, and the general building construction are unique in Britain. The suitability of the flagstones for building has been recognised from prehistoric times – the most famous examples being the Stone Age constructions of Skara Brae and Maeshowe in Orkney. Most older buildings are made of flagstones, usually from local quarries. For example, coastal quarries at South Head, Wick, provided much of the stone with which the town is built. The 'export' of flagstones from Caithness started in 1825 to many cities and towns, such as Torquay, Newcastle, Glasgow and London in the UK, but Caithness flags can also be seen in Hamburg, New York, Melbourne, Calcutta, Bombay, and Rio de Janiero. Crampton and Carrruthers (1914) provide an interesting historical account of the industry. In the last years of the 19th century export ran at around 20,000 tons/annum with a value of around £1/ton and 400–500 people were employed in the industry. Crampton and Carruthers (1914) record that the decline in the industry which set in around 1908 was due to the increasing use of concrete for pavements; thus by 1911 production was down to about 6,000 tons worth only £4,150, with 145 people employed. The industry collapsed after the First World War and virtually all quarries were abandoned, the industry clinging on at Spital prior to recent expansion. The best quality paving slabs were obtained from a number of quarries which included the cliff-top quarry at Ness of Litter near Holborn Head, Spital Quarry and quarries at Castletown where the Castlehill Flagstone Trail provides an insight into the history of the Industry. Lesser quality flags were used for farm buildings and the characteristic Caithness field walls of upright slabs.

Achanarras Quarry was worked for thin micaceous flags that were used as roofing slates. The slates were difficult to 'hole' and slate working had been abandoned by the time Crampton and Carruthers wrote the Caithness Memoir. Evidently the Achanarras fish bed itself was used for slates, since I have seen a *Rhamphodopsi*s from the central part of the fish bed preserved on a roof slate. Slates were also produced from a quarry named Whitemoss, some 3.5 miles SE of Thurso, which would place it near Weydale (Specimens in BGS Collection). Slates were produced until recently on a small scale at Achavrole near Halkirk, particularly for maintenance of historic buildings such as St Magnus Cathedral on Orkney, but the quarry has been infilled with waste from Calder Quarry.

The flagstone industry has revived in recent years, with current (late 2008) traditional working of flagstones at Spital Quarry, and at Achscrabster. Flagstones have been worked recently at Cairnfield Quarry, Weydale, and at Calder Quarry, but these two quarries are currently inactive apart from small-scale working for local use. Increasing use of natural stone in our cities has resulted in considerable output of flagstones in the past 10 years, and examples of recent use can be seen in Glasgow, Edinburgh, Aberdeen, Newcastle and many other places. There is also considerable production of crushed rock for roads and building purposes.

Sand and Gravel

There are numerous small sources of sand and gravel in the area, usually working fluvioglacial material. Beach and blown sand has also been exploited in the past, but destruction of beach and dune systems is no longer acceptable. The sand quarry that existed in the Lower Cretaceous erratic at Leavad produced sand that was used as an abrasive in the machines used to cut the flagstones (Crampton and Carruthers, 1914).

Extensive sand and gravel deposits are present south of Helmsdale on the low-lying coastal platform. Sand and gravel has been worked on a small scale for many years from the low coastal platform near Loth.

Limestone

The importance of lime for agriculture led to use of many of the fish bed laminites as a local source. Examples are the limestones at Port of Brims and at Baligill (Exc. 5), where well-preserved lime kilns can be seen. Shell sand from beaches was also burnt for lime.

Building Stones

A great variety of rocks have been used for the older buildings in the towns of the area. The rock was obtained locally and seldom transported more than a few miles. Thus, Golspie is dominated by red sandstones of the local Old Red Sandstones, and Brora by white to grey sandstones of the Clynelish Quarry Sandstone from Clynelish. Some of the best blocks of Clynelish Quarry Sandstone can be seen in the walls of Dunrobin Castle.

In Caithness numerous local flagstone quarries were used to supply stone for general building work. The variety of building materials, such as sandstones from the Wick Flagstones, the John o' Groats Sandstone and Dunnet Sandstone have all been used, giving subtle differences to the colour and architecture of the towns.

Coal

The earliest reference to coal at Brora dates from 1529 (see Owen, 1995), and Brora Coal was first mined in 1598 when the Countess of Sutherland opened a coal pit. Her son, the fifth Earl of Sutherland, is recorded as re-opening the works in 1634 (Flett *in* Lee, 1925) and four or five pits were sunk, in one of which 15 men were killed by a roof-fall. A John Williams worked the coal for five years starting around 1764. These early pits were near the shore on the south side of the river, and they exploited coal to a depth of about 100 ft ($c.30$ m). The coal was used to evaporate salt water in salt pans built near the pits. 'Salt Street' is a reminder of this industry.

About 1810 a shaft was sunk on the north bank of the Brora River and reached coal at 259 ft (*c.*78 m). From the account quoted by Lee (1925), this appears to be the same shaft that was being used in the 1920s. By this time another pit on the south side of the river was already disused.

In 1956 the N.C.B. estimated reserves of coal at over 12 million tons and suggested testing of the Clynelish and Northern fault blocks with boreholes. The reports of Ewing (1956, 1958) provide information on borings and the structure of the coalfield. In 1966 the Highlands and Islands Development Board put down five boreholes and proved the presence of mineable coal in the inland blocks (Berridge, 1967). An inclined adit was opened in 1969, but the mine ceased production in 1974 and the workings were abandoned in 1975.

The coal seam was generally about a metre thick with a dirt parting in the middle containing a band of pyrite. The coal burnt well, but with an objectionable sulphurous odour, and left a fine white ash that 'every breath of air sent floating over carpets and furniture' (Miller, 1859). Thus, it was not popular with housewives as a house coal. The coal was also prone to spontaneous combustion when exposed to air and damp due to the presence of pyrite. It was the spontaneous combustion of a cargo sent by John Williams to Portsoy which lost him customers who were afraid of such a dangerous cargo! Murchison (1827) recorded a production of 5–6,000 tons/annum from 1814, and in 1910 the production was similarly 6,000 tons. In the 1920s about 30 tons/day were raised with the coal being used in the brickworks and Brora Wool Mill, as well as for domestic consumption. The site of the coal mine has now been landscaped. The Brora Coal is extensive offshore and is present in the Beatrice Oilfield wells, but there are no longer any exposures of the coal in the Brora area. An excellent illustrated history of coal mining at Brora by Owen (1995) provides further information on the industry.

Brick Clay

The Brora Brick Clay of the Brora Argillaceous Formation was formerly dug on the north side of the Brora River for brickmaking, but the pit has now been filled, and other exposures have been landscaped along with the site of the coal mine. Local bricks stamped 'Hunter-Brora' were made from the brick clay, and can sometimes be seen on the foreshore at Brora, generally derived from dumped builders' waste.

Oil

On a clear day the production platforms of the Beatrice Oilfield can be seen from the shore between Brora and Wick. This is the only UK producing oilfield visible from shore, lying only 14 miles offshore from Lybster. The field was discovered in 1976 by Mesa Petroleum with their first exploration well on block 11/30. It was named Beatrice by T. Boone Pickens, the president of Mesa Petroleum, after his wife. Oil was found in Jurassic sandstones ranging from Lower to Upper Jurassic. The main reservoir (the 'A' sandstone) is of Middle to Upper Jurassic age and is very similar in character to the Brora Arenaceous Formation. However, the sandstones are not the same age, the Brora Sandstone being younger than the 'A' Sand. Production also comes from the 'B' sandstones which are the time equivalents of the Brora Roof Bed, and from sandstones equivalent to those in the Brora Coal and Dunrobin Bay formations.

The oil produced is a low sulphur, paraffinic crude with a high wax content (17%) and a high pour point (24°C). It differs from normal North Sea crude, which was largely sourced from organic-rich Kimmeridgian shales, and thus the source of Beatrice oil has given rise to considerable speculation. The Kimmeridge shale is usually immature in the Inner Moray Firth (not buried deeply enough to generate hydrocarbons) and is also not as rich in organic oil-prone material as in areas where it is the source of oil. A source within the Middle Jurassic is possible, particularly the shales of the Inverbrora Member of the Brora Coal Formation which are in part extremely rich in organic material (>25%) and approach the consistency of oil shale. Distillate yields of 32 and 26.5 gallons/ton have been recorded. Although these shales are only marginally mature at Brora, they are more deeply buried to the north and could have provided oil to charge the Beatrice Oilfield by updip migration. Organic geochemical data are indicative of partial sourcing of Beatrice oil from the organic-rich flagstones of the Middle Devonian (Duncan and Hamilton, 1988; Peters *et al.*, 1989), which were deeply buried in the Wick Basin in the Mesozoic (Trewin, 1989).

Production of Beatrice Oilfield started in 1981 with initial reserves calculated at 476 million barrels of oil in place with 162 million recoverable (Linsley *et al.*, 1980). Production peaked at around 50,000 barrels/day in 1984. Ultimate recovery was downgraded to 146 million barrels by 1990 (Stevens, 1991), of which 126 million barrels had been produced by the start of 1992. Despite the prediction in 1992 from BP that the field would not last beyond the Millennium (B.P. pers. comm. January 1992), it is still producing in 2009, and is now operated by Ithaca Energy, who have found new oil-bearing structures in the area. The Polly accumulation lies east of Beatrice within drilling range of an existing Beatrice platform, and Jacky, 10 km NE of Beatrice, started production in April 2009. Ithaca Energy predict combined production from Beatrice and Jacky to be 10,000 barrels/oil/day in 2009.

An onshore exploration well near Lothbeg Point (Sutherland No. 1) in 1980 proved unsuccessful, but small new fields called Lybster and Knockinnon have been discovered close to the Caithness shore. Lybster will be produced from wells drilled from land to the offshore field, the first well being completed in 2008, but there has been no resulting prodution.

The Middle Old Red Sandstone lacustrine deposits, particularly the fish-bed lithology, are rich in oil-prone organic matter (Marshall *et al.*, 1985; Duncan and Hamilton, 1988). The potential of these rocks for hydrocarbon accumulations has been discussed by Trewin (1989) who concluded that oil was unlikely to be found reservoired onshore in Devonian strata, but might be found offshore where Middle ORS is deeply buried by Mesozoic strata, and better possibilities exist for sealing any oil migrated from the Devonian into younger rocks. Solid hydrocarbons are sometimes found associated with vein deposits in the flagstones, the hydrocarbons being derived from maturation of organic matter in the flagstone (Parnell, 1983). Minor mobile hydrocarbons were present in sandstones encountered during drilling at Cairnfield Quarry, Weydale (NHT pers. obs., 2004).

Gold

The short-lived gold rush of 1869 was started by the finding of alluvial gold in the Kildonan and Suisgill burns, which are tributaries of the Helmsdale River (Rice, 2002). Details and references are given in Excursion 6, to which the reader should refer. The gold originates from the migmatised Moine metasediments and has been concentrated by glacial erosion and reworking of the glacial debris by the streams. Deep weathering prior to glaciation may have aided the release of the gold. The original diggings were stopped by the Duke of Sutherland on account of the damage that was being done to the fishing in the Helmsdale by the silt carried from the diggings. (The diggers may also have been partial to salmon and venison.)

Undoubtedly gold remains in the area, washed down into the gravels of the Helmsdale. However, it is unlikely that a short-term alluvial working would benefit the local community, particularly if the salmon fishing was affected. The gold is still panned on a recreational basis with the permission of the estate, and this appears to be the best economic use of the resource at present.

Metallic ores

There are few mineral veins in the area, but a lead/zinc vein was worked on a small scale near Achanarras, and thin veins are sometimes found in the spoil material at Achanarras Quarry. A copper-bearing vein is said to have been worked on the coast south of Old Castle of Wick in the 15th century (Crampton and Carruthers, 1914). Tweedie (1979) reported minor uranium mineralisation from the Helmsdale Granite and also noted Cu–Mo mineralisation in the Ord Burn. Rice (2002) describes the metallic ores from the area.

Non-metallic minerals

A vein of barytes over 6 ft wide at Ray Geo near Lybster has been worked for the mineral, and minor veins of barytes occur throughout the area, emanating from the Helmsdale Granite, which also has small veins of fluorite (Tweedie, 1979). Calcite veins are frequently seen, particularly in fractures adjacent to the Helmsdale Fault (Exc. 3, Itin. 4) and cutting the Middle ORS flagstones.

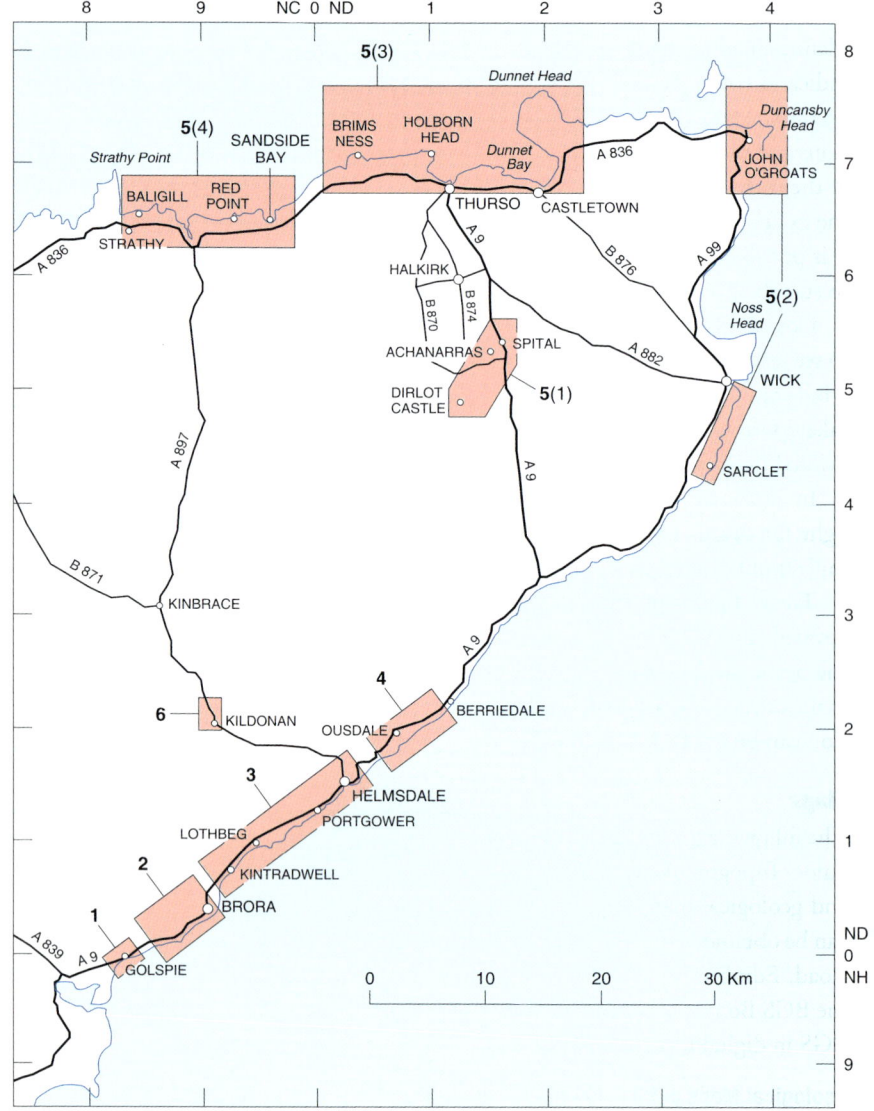

Excursion Localities
1 Golspie; **2** Brora; **3** Kintradwell to Helmsdale; **4** Ousdale; **5** Caithness; **6** Kildonan.

Excursion planner

The localities covered by the excursions in this guide are shown on the Excursion Planner map together with the main roads. A list of the excursions gives a general indication of the time required for each excursion, but the actual time taken depends on the size and speed of the party, and time spent looking for fossils. Weather conditions and the need for 'refreshment' can either extend or shorten time spent looking at the rocks. On the excursion maps 'H' indicates a convenient hostelry! Some of the excursions are subdivided into itineraries and have several access points so that it is possible to plan short routes, or to combine parts of more than one excursion to suit the needs of the party.

Coastal excursions, particularly those on the Jurassic rocks, require half to low tide to see features exposed on the wave-cut platform. In rough weather with an onshore wind some sections may not be safely accessible. At all times great care should be taken; wet grass and rocks in cliff areas are frequently very slippery. Tide times are printed in the local newspaper (Press and Journal) or tide tables can be consulted.

In late autumn and winter lack of daylight becomes a problem, with sufficient light for detailed work only available between 9 am and 3.30 pm. However, at midsummer the days are very long!

The main towns are Golspie, Brora and Helmsdale in the south of the area covered, and Wick and Thurso in the north. Helmsdale and Thurso are considered the best centres at which to stay. Tourist information is available in season in the main towns, or from VisitScotland through visithighlands.com where accomodation can be booked online.

Maps

The following geological and topographic maps cover the localities described in this guide. Topographic maps are generally available from bookshops and tourist offices, and geological maps are stocked by the Orcadian Stone Company in Golspie, or can be obtained from the British Geological Survey, Murchison House, West Mains Road, Edinburgh. Maps and other geological output can be ordered online from the BGS Bookshop, and maps and other information is increasingly available from BGS in digital format.

Geological Maps published by the British Geological Survey (BGS)

BGS   1:625 000 North sheet (covers the whole of Scotland).

BGS   Sheet 58N 04W 1:250 000 Caithness (New series incorporating offshore geology).

BGS   1:50 000 Series
        Sheet 103 Golspie
        Sheet 110 Latheron

Sheet 116E Wick
Sheet 116W Thurso
Sheet 115E Reay
Sheet 115W (Not available)
BGS 1:25 000 Geology Series
Dounreay (bedrock)

Topographic maps
Ordnance Survey 1:50 000 Series
Sheet 10 Strathnaver
Sheet 11 Thurso and Dunbeath
Sheet 12 Thurso and Wick
Sheet 17 Helmsdale and Strath of Kildonan
(There is considerable overlap between sheets 11 and 12, sheet 12 being the best value for Excursion 5.)

Major topics covered by excursion guides
In the following list the main features covered by each excursion are presented in note form to enable the reader to scan the list for items of particular interest. This list can be used in conjunction with the locality map and list of times required for excursions. Composite excursions can then be constructed to cater for the interests of the party within the time available.

1. **Trias to Lias**
2. **Brora Coal Formation to Balintore Formation**
    - Itin 1   Brora Coal to Brora Arenaceous Fm. Brora Foreshore.
    - Itin 2   Brora Arenaceous Fm. S. bank of Brora River.
    - Itin 3   Brora Argillaceous Fm. N. bank of Brora River.
    - Itin 4   Balintore Fm. N. foreshore at Brora.
3. **Upper Jurassic sedimentation controlled by the Helmsdale Fault**
    - Itin 1   Kintradwell Boulder Beds, soft sediment deformation.
    - Itin 2   Sandy submarine 'turbidite fan' deposits, Allt na Cuile Sst.
    - Itin 3   'The Fallen Stack'; boulder beds and clast stratigraphy.
    - Itin 4   Helmsdale Boulder Beds; Helmsdale Fault Zone.
4. **Lower ORS and relation to Helmsdale Granite**
5. **Old Red Sandstone of Caithness**
    - Itin 1   Achanarras fish bed at Achanarras Quarry, lacustrine laminites, lacustrine margin at Dirlot.
    - Itin 2   John o' Groats Sandstone, Wick Flagstones at Wick, Sarclet Group at Sarclet Haven.
    - Itin 3   Middle ORS cyclicity. Brims Ness, Holburn Head fish bed, Pennyland Shore at Thurso, Dunnet Sandstone.

	Itin 4	Basin margin deposits at Red Point, Port Skerra and Baligill; ORS on Moine basement. Sandside Bay, marginal lacustrine and aeolian.
6.	**Gold panning. Moine basement**	

Guide to times required for excursions

The times quoted do not include driving times to the starting point. Times allow for discussion and close examination of the rocks – a quick 'look-see' excursion will take less time.

1. **Golspie, Triassic–Lower Jurassic**
   2–3 hours. Low tide required.
2. **Brora, Middle–Upper Jurassic**
   Itin 1   3 hours. Low tide required for Locs 1, 2
   Itin 2   3 hours
   Itin 3   3 hours
   Itin 4   1 hour. Low tide required
3. **Kintradwell–Helmsdale Upper Jurassic**
   Itin 1   2 hours. Low tide required
   Itin 2   4 hours for full excursion. Low tide required for localities 3, 5, 6.
   Itin 3   2 hours. Low tide required
   Itin 4   1–2 hours for localities 1–4; half day for localities 1–9. Low tide required for all localities.
4. **Ousdale Lower ORS**
   2 hours. Can be done en route to Exc. 5, Itin. 1 from Helmsdale.
5. **Caithness Old Red Sandstone**
   Itin 1   Achanarras, Dirlot. Half day minimum, but much longer if searching for fossil fish.
   Itin 2   John o' Groats, Wick Sarclet. 1 Day. Low tide required at John o' Groats for localities 6, 7, 8.
   Itin 3   Brims Ness to Dunnet Head. 1 Day. Low tide required for Brims Ness (Loc. 11); Pennyland Shore, Thurso (Loc. 13) and Clett Harbour (Loc. 15).
   Itin 4   Strathy to Red Point and Sandside Bay 1 Day. Low to half tide advisable for Port Skerra (Loc. 18) and Sandside Bay.
6. **Kildonan Gold**
   Half day minimum. Depends on time spent gold panning.

Brief highlights and potential problems

For those visiting the area for the first time, and those with limited time at their disposal, the following might be considered the highlights of the area, and the associated problems.

Jurassic

Lias at Golspie (Exc. 1)
Poor exposure due to sand cover in intertidal area.

Brora Coal to Balintore Fms. (Exc. 2)
Itinerary 1 and Itinerary 4 cover most of the succession, exposure is mainly on the foreshore amongst boulders, hence there are problems with seaweed and sand cover. River exposures (Itins. 2, 3) are particularly useful if tides are too high.

Upper Jurassic fault controlled sedimentation (Exc. 3)
Kintradwell Boulder Beds. Itin. 1.
Sandy delta-derived 'fan' deposits of Allt na Cuile Sst. Itin. 2, Locs. 4–6.
'Fallen stack' at Portgower. Itin. 3, Loc. 4.
Helmsdale Boulder Beds. Itin. 4, Locs 1–4.
All shore exposures are subject to variable sand cover and seaweed in late summer.

Devonian (Excs 4 and 5)

Lower ORS. Ousdale arkose and Ousdale Mudstones. Exc. 4. Sarclet Group. Exc. 5, Itin. 2.
Achanarras fish bed, Achanarras Quarry. Exc. 5, Itin. 1.
Lacustrine margin, stromatolite coated breccia, Dirlot Castle. Exc. 5, Itin. 1.
Wick Flagstones sedimentology and cycles, South Head Wick. Exc. 5, Itin. 2.
Middle ORS cycles. Brims Ness and Thurso Shore. Exc. 5, Itin. 3. Low tide needed.
Upper ORS fluvial deposits, Dunnet Head. Exc. 5. Itin. 3.
Lacustrine margin unconformity at Red Point and Port Skerra. Exc. 5. Itin. 4.
Aeolian sandstones and flagstone cyclicity, Sandside Bay. Exc. 5 Itin. 4.

Basement and mineralisation

Gold at Kildonan. Exc. 6. (Problem, not a lot of gold)
Basement Geology. Moines and intrusives. Exc. 6.
Basement also seen at Baligill, Port Skerra and Red Point. Exc. 5, Itin. 4.

Excursion 1

The Triassic and Lower Jurassic of Golspie

N. H. Trewin

Purpose
To examine the Triassic and Lower Jurassic (Lias) of the Golspie to Dunrobin shore section.

Access
The exposures can be reached by walking from Golspie or by combining the excursion with a visit to Dunrobin Castle. If starting from Golspie the workshops and excellent geological exhibition of the Orcadian Stone Company in Main Street can be visited. At the east end of Main Street the A9 bears sharp left into Old Bank Road; at this corner turn right into Duke Street in which there is parking space. Cross the Golspie Burn footbridge, follow the track into the fields, and walk about 1 km east to locality 1 which is near the low tide mark beyond the remains of an old stone pier and wood piling. If visiting Dunrobin Castle, a track leads to the shore down the hill past the western side of the castle and reaches the shore at locality 3 (Fig. 1.1). Low tide is essential for this excursion, which takes 2–3 hours. In recent years exposure on the beach has been poor due to sand and seaweed cover, and a good winter storm is needed to reveal much of the geology.

Introduction
The general geology and regional features of the Triassic and Lower Jurassic strata are described in the Geological History section of this volume. For a broad account of Triassic rock in the Moray Firth see Glennie (2002), and for the Jurassic see Hudson and Trewin (2002).

Judd (1873) provided an early description of the section, and considerable detail of Jurassic faunas is given by Lee (1925). The geological succession is illustrated in Figure 1.2. The Triassic is poorly exposed, but includes fluvial and aeolian yellow sandstones overlain by mudstones with extensive caliche development and some silcrete. This palaeosol horizon correlates stratigraphically with the similar 'Cherty Rock' of Lossiemouth.

The Jurassic section comprises the Dunrobin Bay Formation of Batten *et al.* (1986). Richards *et al.* (1993) modified the stratigraphic nomenclature (Fig. 1.2), raising the Dunrobin Bay Formation to Group status and introducing new formation names. The descriptive terminology of Batten *et al.* (1986) is followed here.

The Dunrobin Pier Conglomerate Member of alluvial origin was discussed by Batten *et al.* (1986). The overlying Dunrobin Castle Member was described by Neves and Selley (1975) on the basis of borehole information, and they demonstrated that the lower, unexposed part of the Lias consists of shales and siltstones

1.1 Locality map, excursion 1.

with occasional thin coals and rootlet beds. These strata were deposited in a dominantly coastal alluvial environment; however, some horizons containing dinocysts are indicative of marine influence (Lam and Porter, 1977). The White Sandstone Unit represents a coastal sand body and is overlain by shallow marine strata of the Lady's Walk Shale Member. The numbered bed-by-bed succession described by Lee (1925) cannot usually be demonstrated due to poor exposure, but some of the more prominent beds are usually visible at low tide and yield a varied fauna dominated by bivalves. The uppermost strata, consisting mainly of shales with calcareous nodules and some sandstone beds with a few pebbles and a shelly fauna, were possibly not exposed at the time of Lee's work.

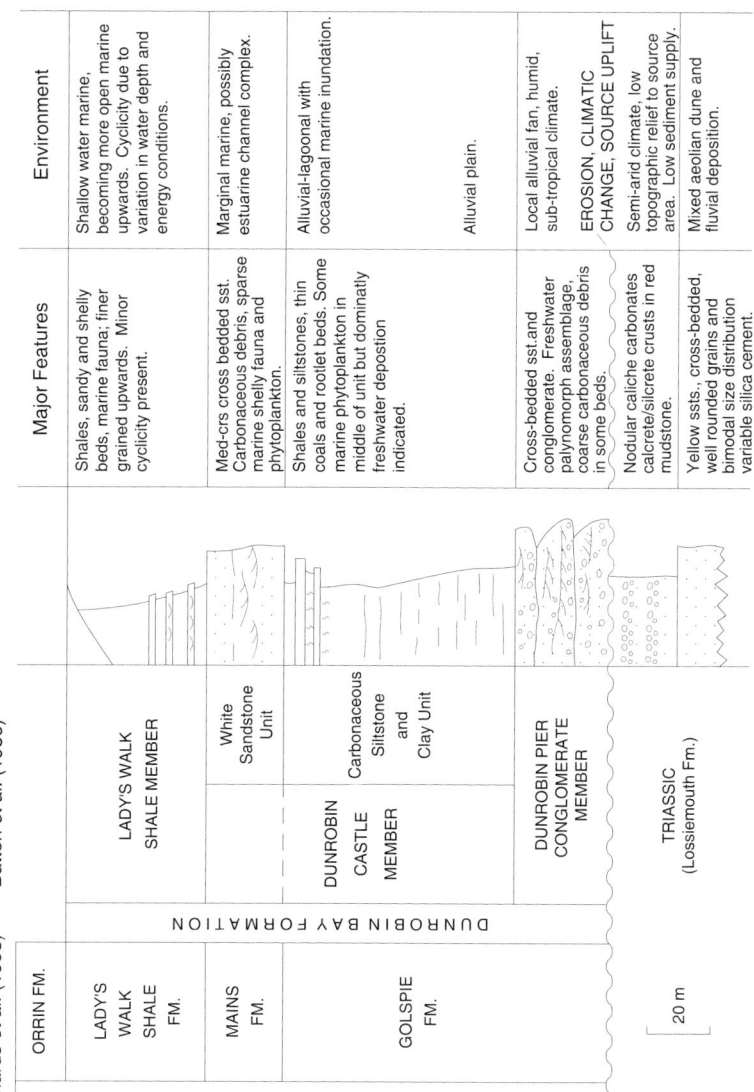

1.2 Stratigraphy, lithofacies and environmental interpretation of Triassic and Liassic strata at Golspie (Modified from Batten et al. 1986).

Locality 1. Triassic Sandstones [NC 849 003]

The lowest sandstones seen here are yellow, laminated, and have a bimodal grain size distribution with well-rounded and spherical 'millet seed' medium sand grains in a fine sand matrix. Lamination is a reflection of grain size and sorting variation. The strike of these sandstones varies between 190° and 280°, with dips to the west and north.

The overlying sandstones contain irregular concretionary patches with a siliceous cement and dip NE (035°) at 15–20°. It is possible that the variable dip and strike of the underlying sandstones represents poorly preserved, large-scale cross-bedding. These sandstones may be broadly equivalent to the aeolian Lossiemouth Formation of the Triassic of the Elgin area (Peacock et al., 1968; Gillen, 1987).

Locality 2. [NC 850 004]

The sandstones of locality 1 are overlain by poorly exposed red and green mudstones and marls (calcareous mudstones). At locality 2 extensive developments of concretionary carbonate are present in the mudstones. The carbonates are irregular concretions of pink to grey micritic limestones with white calcite veins. These limestones are caliche type carbonates formed in the soil profile of a semi-arid region where evaporation exceeded precipitation. Some examples of calcrete textures may be seen with multiple carbonate veining, brecciation features, altered clasts of country rock (mainly sandstone) and occasional carbonate pisoliths (Fig. 1.3). Some of these rocks have been partly silicified to form a silcrete.

1.3 Cut section of caliche limestone from the top of the Triassic section. Specimen from Golspie Glen.

These calcrete- and silcrete-textured rocks have also been seen in the Golspie Burn near the rail bridge, and fragments noted some 300 m upstream of the bridge; thus, these rocks form an extensive unit near the top of the Trias. They can be correlated with the Stotfield Cherty Rock at the top of the Trias of the Elgin area (Phemister, 1960; Gillen, 1987) and form an extensive seismic marker horizon in the Inner Moray Firth (Andrews and Brown, 1987).

Locality 3. Dunrobin Pier [NC 8515 0050]

The Dunrobin Pier Conglomerate is now taken as the basal member of the Dunrobin Bay Formation (Batten *et al.*, 1986, Fig. 2). Exposures are now more extensive than in the past due to the destruction of Dunrobin Pier by storms. A diagrammatic section of the main exposure (sand permitting) in the small bay is given in Figure 1.4. The three calcite-cemented units form reefs on the shore, but the poorly cemented sandstones and shales are rarely exposed. The prominent reefs on

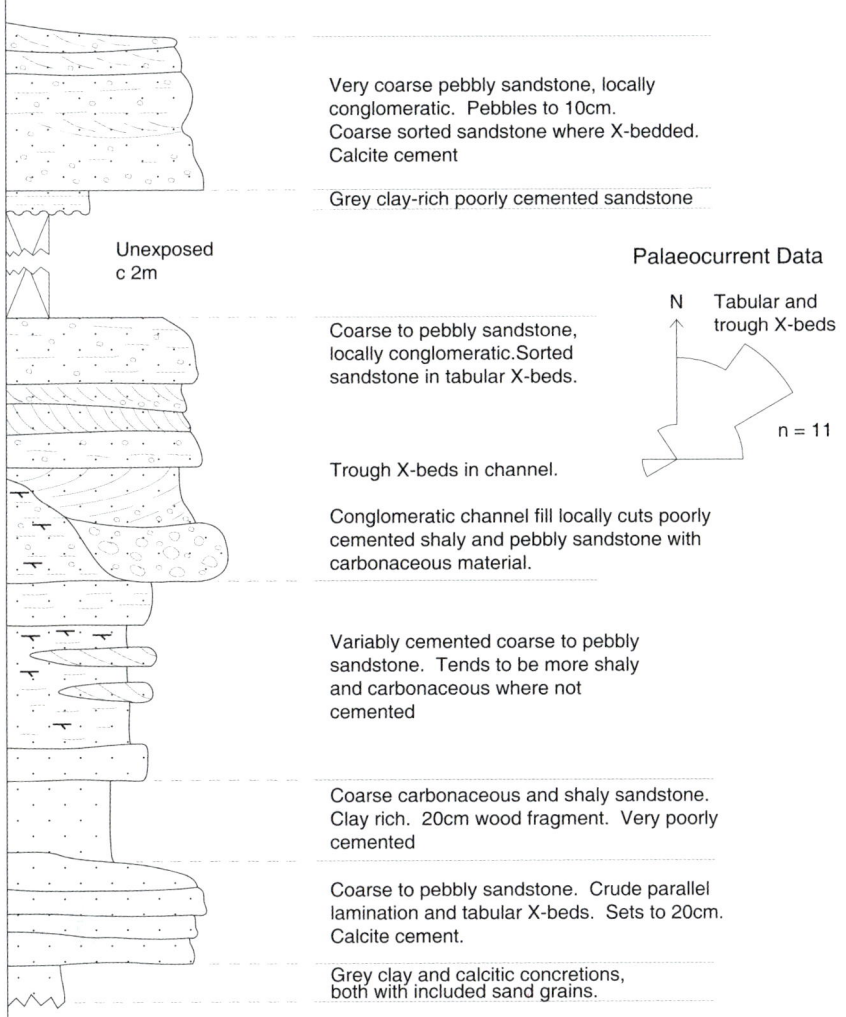

1.4 Log of the Dunrobin Pier Conglomerate Member (Modified from Batten *et al.* 1986).

the shore consist of coarse sandstone and conglomerate with a poikilotopic calcite cement. About 90% of clasts are of sandstones, limestones and cherty rocks identical to underlying Triassic lithologies, and were clearly derived by local erosion of the Triassic. Other clasts are of vein quartz and metamorphic quartzite. It is probable that the Dunrobin Pier Conglomerate rests with an erosive contact on the underlying Triassic. The rapid local derivation of the Dunrobin Pier Conglomerate could be due to early Jurassic basin margin faulting uplifting Triassic strata and causing local erosion, possibly by movement on the Helmsdale Fault.

The conglomerate has been reported as unfossiliferous (Judd, 1873; Lee, 1925; Neves and Selley, 1975), but temporary exposures following a storm that stripped sand from the beach revealed carbonaceous shales and sandstones interbedded with the conglomerates. The shales yielded a rich terrestrial micro-flora and woody debris up to 20 cm long (Batten *et al.*, 1986). There is no evidence of marine forms in the assemblages.

The conglomerates and sandstones were deposited under high-energy fluvial conditions, probably in braided channels on a small alluvial fan. Evidence from cross-bedding indicates transport to the NE (Fig. 1.4). A marked climatic change at the end of the Triassic resulted in increased humidity and consequent vegetation. The age of the conglomerate is probably early Hettangian, but a Rhaetian age is not excluded (Batten *et al.*, 1986).

Judd (1873) estimated the thickness of the conglomerate as 50 ft(c.15 m), a figure with which Hurst (1985) agreed. Lee (1925) mentioned that only 6 ft was exposed at the end of the pier; this seems to be the bed described as 2 m thick by Neves and Selley (1975), a thickness that has unfortunately been commonly repeated in the literature. Although the section is disturbed by faults it appears that at least 32 m of conglomerate and sandstone are present (Batten et al., 1986).

Locality 4. The White Sandstone [NC 856 009]

From locality 3 walk along the beach to the first prominent reef which is formed by the White Sandstone Unit of the Dunrobin Castle Member (Fig. 1.2). The underlying Carbonaceous Siltstone and Clay Unit is rarely exposed. At locality 4 about 11 m of white, medium- to coarse- grained sandstone is exposed. Neves and Selley (1975) proved a further 12 m by drilling. The sandstone is cross-bedded and contains carbonaceous debris and thin shaly partings. Lee (1925) recorded marine bivalves including *Grammatodon* from clays some 5 m above the main sandstone, and Neves and Selley (1975) reported marine microplankton from the sandstone.

This sandstone, lying between dominantly non-marine alluvial coastal plain deposits and the shallow-water marine deposits of the Lady's Walk Shale Member, represents a coastal or estuarine sand body. It may represent part of an estuarine channel complex, or sandy barrier bar. Exposure is too poor for detailed interpretation.

The coarse sandstone has excellent porosity (23%) and permeability (1600 mD horizontal and 880 mD vertical measured on core plugs) and has some reservoir potential. Minor production in the Beatrice Oilfield (Linsley *et al.*, 1980; Stevens, 1991) is from broadly equivalent sandstones of Lower Jurassic age, but the sandstones of this age in the Beatrice Field have much lower porosity and permeability due to quartz cement and greater burial depth.

Locality 5. Lady's Walk Shales [NC 858 009]

Between localities 4 and 5 there is usually variable exposure of the Lady's Walk Shale Member. The sequence is dominated by shales with calcareous nodules, but includes thin limestones, sandstones and bioturbated units with shelly faunas. Several beds contain coarse sand and pebbles.

The numbered bed sequence of Lee (1925) cannot usually be verified and the thickness given by Lee (18 m+) seems too low, there being more than 32 m exposed, and a probable total of 48 m+ (Batten *et al.*, 1986)

The exposed reefs of harder rock comprise calcareous sandstones, some with a rich bivalve fauna dominated by small *Ostrea*, and including *Gryphaea*, *Modiola*, *Pleuromya* and pectinids. Rhynchonellid brachiopods are also present. The harder beds are parts of small-scale coarsening-up units that start with micaceous shales with calcareous concretions and pass up into bioturbated sandstones in which some trace of ripple lamination may be preserved. Burrows present include *Rhizocorallium*, *Spongeliomorpha*, *Siphonites* and *Chondrites*. The tops of some beds contain scattered quartz pebbles. The cycles represent decreasing water depth and increase in wave energy, which resulted in winnowing of sands, and probably periods of non-deposition before further deepening initiated the next cycle. The presence of land-derived plant debris in most of the section attests to the closeness of the Scottish landmass.

The highest (stratigraphic) exposures occur in a low cliff at locality 5 and consist of dark blue-grey shales with calcareous concretions, and include several cemented sandstone beds with pebbles, belemnites, rhynchonellids, bivalves and wood debris. These sandstone beds are up to 40 cm thick, laterally impersistent, and appear to have been deposited in shallow channels or erosional gutters, probably at times of lower sea level. Subsequent bioturbation has resulted in mixing of sand, pebbles and bioclastic debris. Variations in the clay mineralogy of the shales are also believed to be indicative of local changes in sea level (Hurst, 1985). Rare ammonites from the Lady's Walk Shale are of Sinemurian (*raricostatum* Zone) and Pliensbachian (*jamesoni* Zone) age. The rarity of ammonites is a reflection of the nearshore environment of the sequence. The regional aspects of this Lias succession are discussed in the Geological History section, where the sequence is compared with the similar development in Beatrice Oilfield (Linsley *et al.*, 1980; Stevens, 1991).

It is worthwhile to examine the many boulders on the beach, since a great variety of rock types and geological phenomena can be observed. Most of the boulders come from local boulder clays or were used in coastal defences which are now in a poor condition. Rock types present include ORS conglomerates and sandstones in which good examples of cross bedding and ripple lamination can be seen. Boulders of both foliated and non-foliated granites are common, and xenoliths of partly digested country rock are present in some. Gneissose boulders exhibit excellent folds and several phases of veins, both of quartz and granitic material. Quartzites and amphibolites from the Moines are also common.

From locality 5 a coastal footpath leads back to the track to Dunrobin Castle and on to Golspie. It is about 25 minutes walking from locality 5 back to the parking area at Duke Street.

Excursion 2

Bathonian to Oxfordian strata of the Brora area

A. Hurst

Purpose

To demonstrate the main features of the succession from the Brora Coal Formation to the Balintore Formation in the Brora area. The excursion is divided into four itineraries that can be timed to make best use of the tides.

Itinerary 2.1 The Brora Coal to Brora Arenaceous formations on south foreshore Brora and at Strathsteven.

Itinerary 2.2 The Brora Arenaceous Formation, south bank of the River Brora.

Itinerary 2.3 The Brora Argillaceous Formation, north bank of the River Brora.

Itinerary 2.4 The Brora Arenaceous and Balintore formations, north foreshore, Brora.

Access and general information

About two days of excursion are described in the four itineraries. The localities are easily accessible on foot, generally along recognised rights of way, and short distances from parking areas. Brora is an excellent starting point for visiting all localities, and has shops, hotels, parking space and public conveniences.

Localities are shown on the general map (Fig. 2.1). The order in which itineraries are taken will depend on the state of the tide; coastal localities 1, 2, 6, 9 and 10 require low tide. Although the tidal range at Brora is considerable (>2 m) there is little danger of being cut off by the rising tide. Should one be surrounded suddenly by the sea, no more should be necessary than a knee-deep wade across the shallow slope of the wave-cut platform. A more hazardous feature of visiting the inter-tidal localities is the seaweed and slimy green algae that often make conditions treacherous underfoot. The shore exposures are cleanest in spring following winter storms. Access details are given in each itinerary.

Introduction

The general geology of the succession is described in the Geological History section of this volume and the stratigraphy summarised in Figure 2.2. Many different workers have attempted to unravel the stratigraphy of the area and measure thicknesses of the units. Poor exposure, shallow dips and probable local thickness variations make this task difficult; hence, thickness estimates vary considerably between authors.

The Brora Coal Formation consists of an essentially fluviatile sequence of channel sandstones and floodplain mudstones of the Doll Member (Hurst, 1981), over-

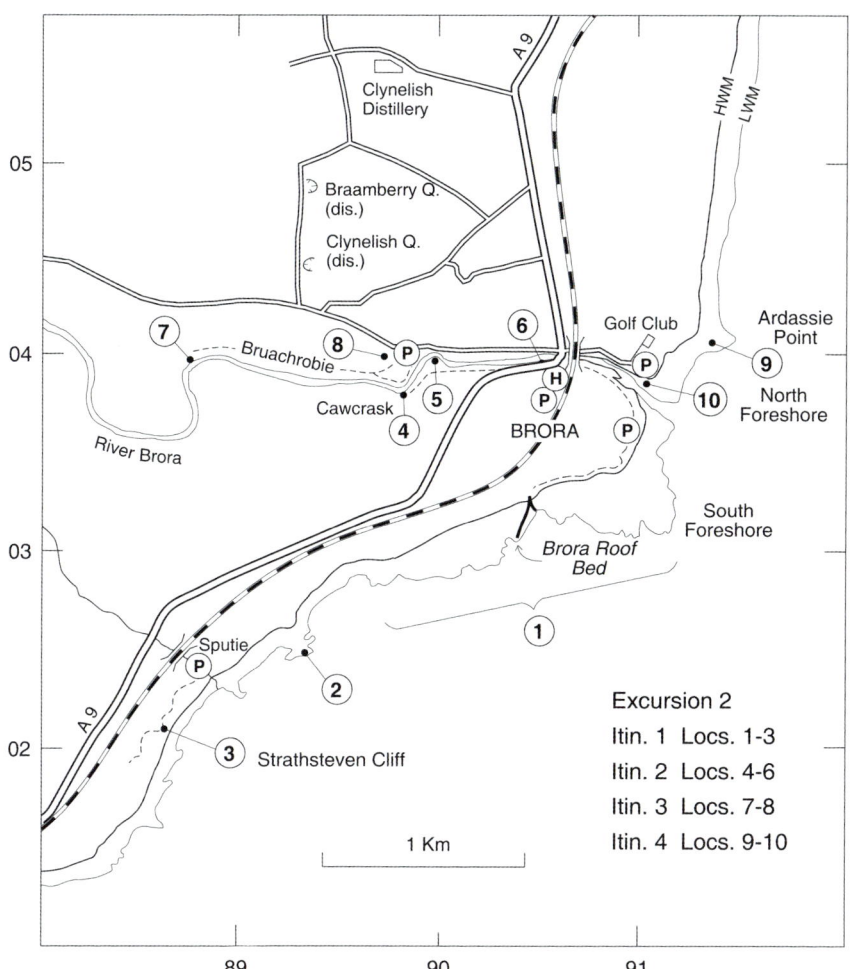

2.1 General locality map of the Brora area.

lain by the dark shales with thin coals, bituminous shales and two shell beds of the Inverbrora Member, which was deposited in a lagoon with variable marine influence (MacLennan and Trewin, 1989). The Brora Coal was deposited during a period of isolation of the lagoonal area from the sea; waterlogged plant debris first accumulated, followed by coal formation over a wide area (as far as Balintore and the Beatrice Oilfield). The Brora Coal was mined intermittently from early times (1598) but the mining finally ceased in 1974 (Owen, 1995). The Brora Coal Formation is usually considered to be Bathonian in age but dinocysts indicate that the base of the Callovian may lie within the Inverbrora Member (MacLennan and Trewin, 1989).

Overlying the Brora Coal is the Brora Roof Bed, a marine sandstone that records a rapid marine transgression and subsequent deepening water, in which the Brora Shale Member of the Brora Argillaceous Formation was deposited. Parts of the Glauconitic Sandstone and Brora Brick Clay members are present on the shore, but are usually poorly exposed and are best seen inland (Itin. 3). Since the comple-

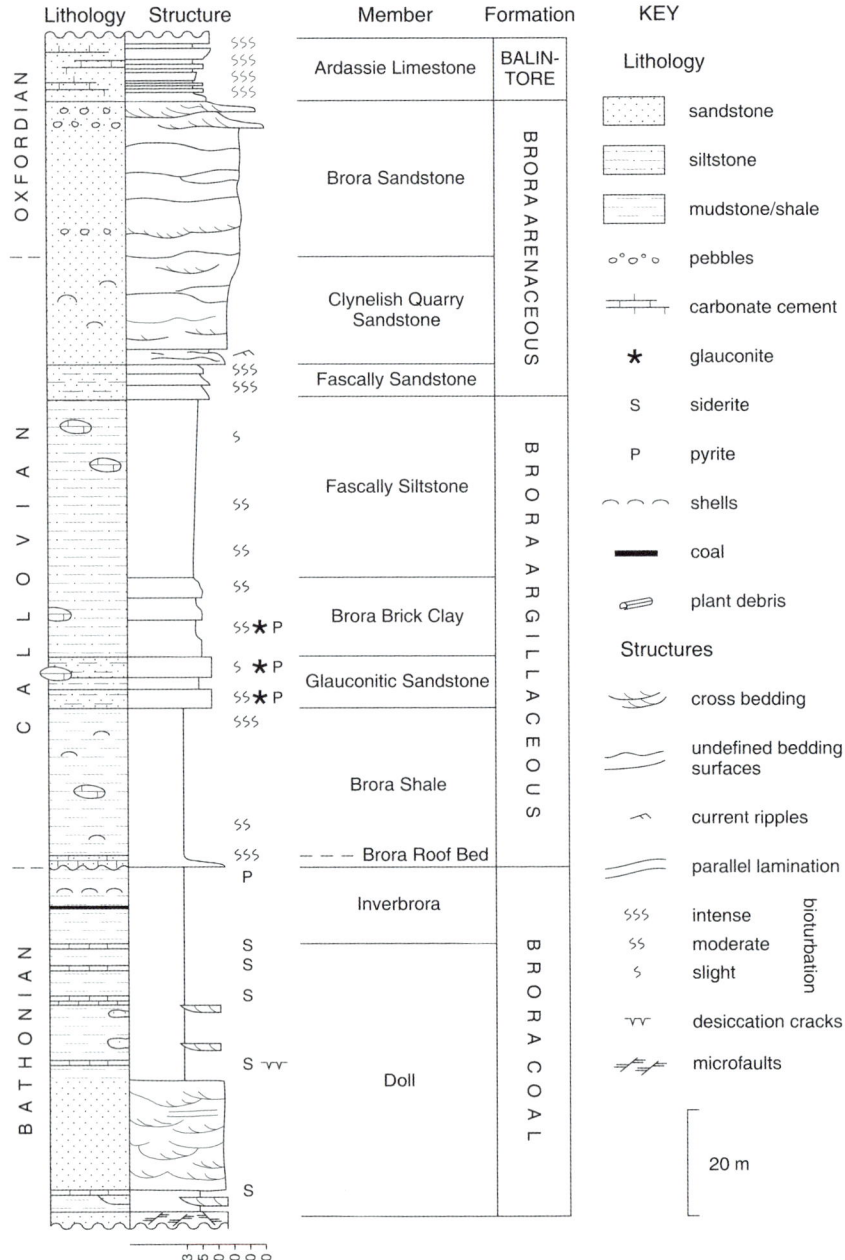

2.2 Stratigraphy and sedimentological log of the Bathonian to Oxfordian section at Brora.

tion of the work by Sykes (1975a, b) many exposures of the Brora Argillaceous Formation have been lost, notably those near the former brick works, which are now landscaped.

The Brora Arenaceous Formation is sandstone-rich, in which a general coarsening-upward character represents deposition in a coastal sandbar environment (Sykes,

1975a). Much of the building stone used in Brora is from the Brora Arenaceous Formation, largely from Clynelish Quarry [NC 893 045] and Braamberry Quarry [NC 893 049]. Ammonites from Clynelish Quarry decorate the clock tower in the centre of Brora.

Very limited exposure of the Balintore Formation is found at Brora, consisting of approximately 12 m of interbedded muddy carbonaceous sandstones and 'limestones', which are siliceous spiculites subsequently recrystallised to carbonate (Sykes, 1975a). Outcrop is restricted to a small area on the foreshore north of Brora [NC 914 041].

The structural dip in the Brora area is generally 10° or less. Steeper dips are recorded near the Brora Fault and along the limbs of the anticlinal structures at Fascally [NC 899 040]. Gentle S-folding sub-parallel to the coastline can be traced, similar to the structural trend identified further NE in the Kimmeridgian Helmsdale Boulder Beds (Exc. 3). Poor exposure limits comparison between the different fault compartments but structural trends are not readily mappable across fault planes, evidence which may be indicative of some lateral movement on faults.

Most of the lithostratigraphic boundaries shown in Figure 2.2 can be mapped reliably, but on a finer scale there are considerable problems when defining the thicknesses and distribution of individual members. From careful examination of Sykes (1975 a, b), all available borehole data and my own field measurements, it is assumed that the thicknesses given in Figure 2.2 are reliable for members of the Brora Coal and Brora Argillaceous formations, but much less reliable for the Brora Arenaceous Formation. Previous estimates of the total thickness of the Brora Arenaceous Formation vary from 30 m (Lee, 1925), to >60 m (Sykes, 1975a), and up to 122 m (Judd, 1873). My preference is for Lee's estimate, structural and thickness measurements making it difficult to accommodate more than a 40 m total thickness for the whole of the Brora Arenaceous Formation.

ITINERARY 2.1
Brora Coal to Brora Arenaceous formations on south foreshore, Brora and at Strathsteven

Purpose
To examine the Brora Coal Formation demonstrating the upward transition from alluvial plain (Doll Member) to marine-influenced lagoonal conditions (Inverbrora Member) and the major Callovian marine transgression that initiated deposition of the Brora Argillaceous Formation. Marine-bar sandstones of the Brora Arenaceous Formation are seen at Strathsteven and the Brora Fault can be demonstrated.

Access
The itinerary can be commenced at Brora or from the small parking area near Sputie [NC 8880 0245]. The track to the parking area at Sputie leaves the main road at [NC 885 025] is narrow and passes under the railway, the tunnel being only just wide enough for a minibus. There is only room for three or four cars at the parking area. This parking can be used for localities 2 and 3 if the party does not wish to walk from Brora.

Starting at Brora either walk to the shore or take Harbour Road by turning east immediately south of the road bridge in Brora. Park at the car park near the old radio station (now a storage facility) and walk south to the point where the sewage pipe crosses the beach (Fig. 2.3). Localities 1 and 2 require low tide, and sand and seaweed cover frequently obscures large areas of outcrop.

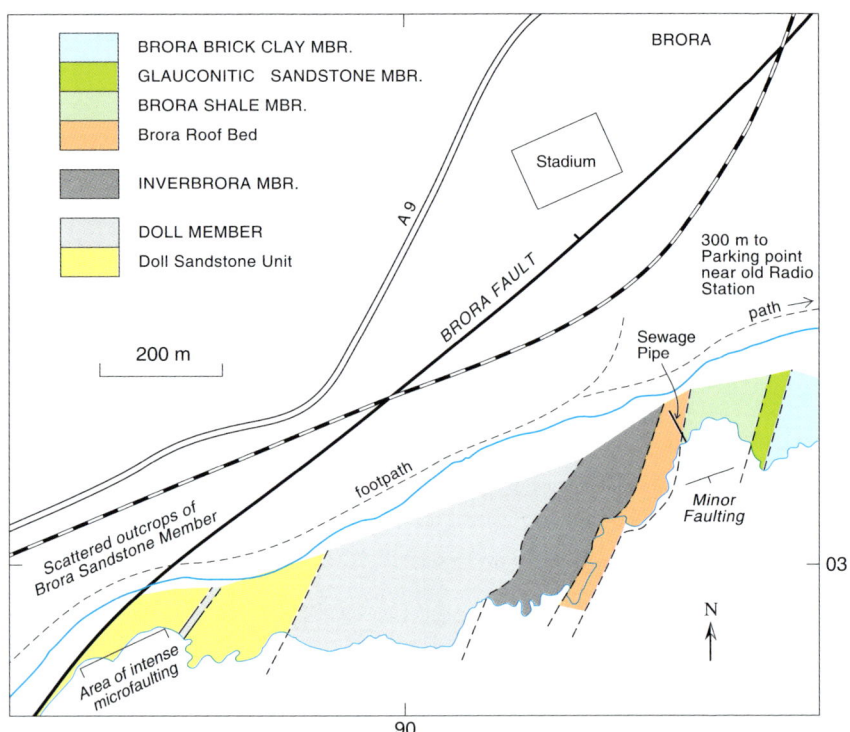

2.3 Geological sketch-map of shore at locality 1.

Locality 1. Brora Coal and Brora Argillaceous formations, south foreshore [NC 905 032]

The excursion is best started from the most easily identified point in the succession, the Brora Roof Bed, which forms a prominent reef marking the seaward limit of exposure along almost 1 km of the foreshore [NC 905 032 to 902 029]. The sewage pipe crosses the beach [NC 905 032] (Fig. 2.3) about 500 m SW of the old radio station, and is fixed to the Brora Roof Bed.

Brora Argillaceous Formation

The Brora Roof Bed is the basal bed of the Callovian Brora Argillaceous Formation (Sykes, 1975a), and was deposited during the regional Lower Callovian marine transgression. Sykes (1975a) describes the Roof Bed as an overall fining-upward sequence of intensely bioturbated medium-grained sandstone with a few quartzite pebbles. The thickness of this bed in the Brora area varies up to 2.3 m.

Fossils are common but difficult to extract from the calcite-cemented sandstone. In some areas on the top surface of the Roof Bed the gastropod *Piettei*a shows a preferred orientation of shell apices towards the SW (Sykes, 1975a), and the bivalves *Myophorella*, *Gervillella*, *Corbula* and *Pleuromya* may be seen, the latter in vertical burrowing position. Reworked fragments of the coal are present.

The sandstone was deposited in a shallow marine coastal setting and rests on the Brora Coal (not exposed). To the south (30 km) at Balintore (Cadh' an Righ) a thinner (0.5 m) laterally equivalent sandstone rests on an eroded and bored surface of the Brora Coal (Sykes, 1975a; MacLennan and Trewin, 1989).

The Roof Bed grades upward rapidly into the Brora Shale (only exposed at low spring tide) that was deposited in an open marine environment, as is evidenced by the presence of ammonites and belemnites. Although the shales are very dark and organic rich, there is a fauna of bivalves present including *Trautscholdia*, *Thracia*, *Protocardia* and *Meleagrinella*. Both shallow- and deep-burrowing forms are present; thus bottom-water conditions were not continuously anoxic. However, reducing conditions prevailed beneath the sediment/water interface as shown by the high organic content. Abundant angular shell debris is probably the result of predation by fish or arthropods on the molluscan population. Both land-derived material (plant fragments and spores) and marine microplankton (dinocysts) are abundant (MacLennan and Trewin, 1989) and are indicative of nearshore open-marine conditions.

About 100 m east of the sewer pipe the first outcrops of the Glauconitic Sandstone Member are seen low on the beach. The typical lithology is a muddy, glauconitic, bioturbated, very fine-grained sandstone with glauconite concentrations in burrows. Sideritic and phosphatic concretions are present and belemnites are abundant in several beds.

Further west parts of the Brora Brick Clay and Fascally Siltstone members form a broad low intertidal platform. Few details are visible here apart from some large calcareous concretions. Toward the river mouth, there are extensive but low-lying exposures of parts of the Brora Arenaceous Formation. Fracturing and hardening of some sandstone outcrops is associated with faulting near the mouth of the Brora River. The Brora Argillaceous Formation is described in more detail in Itinerary 3.

Brora Coal Formation
Return to the starting point at the Roof Bed and locate exposures of dark micaceous and carbonaceous shales of the Inverbrora Member of the Brora Coal Formation due west of the Roof Bed. The Inverbrora Member includes some bituminous shales approaching oil-shale in composition (up to 26% Total Organic Carbon), and two thin green shell beds with *Neomiodon* and *Isognomon*, which are still aragonitic, and occur about a metre apart in the sequence (Fig. 2.4. Abundant drifted plant material is present in the shales, Stopes (1907) having described three species of *Equisetum* and nine other plants including leaves of *Ginkgo*. The shales are mostly finely laminated, contain thin lenses of coal and have small pyrite concretions that sometimes replace plant debris. The shell beds are reworked and winnowed at their tops, with rounded shell debris. Lam and Porter (1977) first recorded marine microplankton from the Inverbrora Member, which has been confirmed by MacLennan and Trewin (1989) who have shown that dinocyst assemblages are present that represent variable

2.4 Shell bed with Neomiodon and Isognomon near the top of the Inverbrora Member. Lens cap 52mm.

marine influence in a lagoonal environment. The presence of marine benthonic foraminifera confirms the influence of marine conditions.

The Brora Coal was deposited when the lagoon became isolated from the sea and abundant plant material including *Equisetum* and conifer wood (Harris and Rest, 1966) accumulated. The absence of a seat-earth below the coal is evidence that the water depth in the lagoon was initially too great to allow rooting of plants; the initial deposits of coal comprise drifted materials. Rootlets have been recorded from a dirt bed within the coal.

The composite section (Fig. 2. 5) of the Brora Coal Formation shows the general features of the outcrop that lies to the west of the Roof Bed. Exposures are low-lying and subject to sand, seaweed and boulder cover; thus, beds described here may not be visible. Below the shell beds the Inverbrora Member becomes lighter grey, less rich in carbonaceous material, and the shales give way to mudstones. This trend continues into the underlying Doll Member where grey mudstone predominates. The top of the Doll Member is marked by the occurrence of a laterally extensive siderite-cemented, grey, brecciated mudstone with a distinctive red-brown weathering colour (Bed 1). Thin rippled sandstones with plant debris are associated with Bed 1 that, when exposed, forms a ridge on the foreshore some 10–20 cm higher than the surrounding mudstones. Finding Bed 1 is usually quite simple and provides a useful line of reference from which the rest of the section can be examined.

Neves and Selley (1975) record the presence of the freshwater bivalve *Unio* from Bed 1 and a rich freshwater ostracod fauna was recovered between Beds 3 and 4 (R. Titterton, pers. comm.) (Fig. 2.5). A rich non-marine palynoflora has been recovered from mudstones of the Doll Member (J. Fenton, pers. comm.) and silicified logs are common along a horizon between Beds 1 and 2 (Hurst, 1981).

Siderite-cemented mudstone horizons ('cementstones' of Lee (1925)) can be correlated laterally and allow the vertical sequence of the Doll Member to be estab-

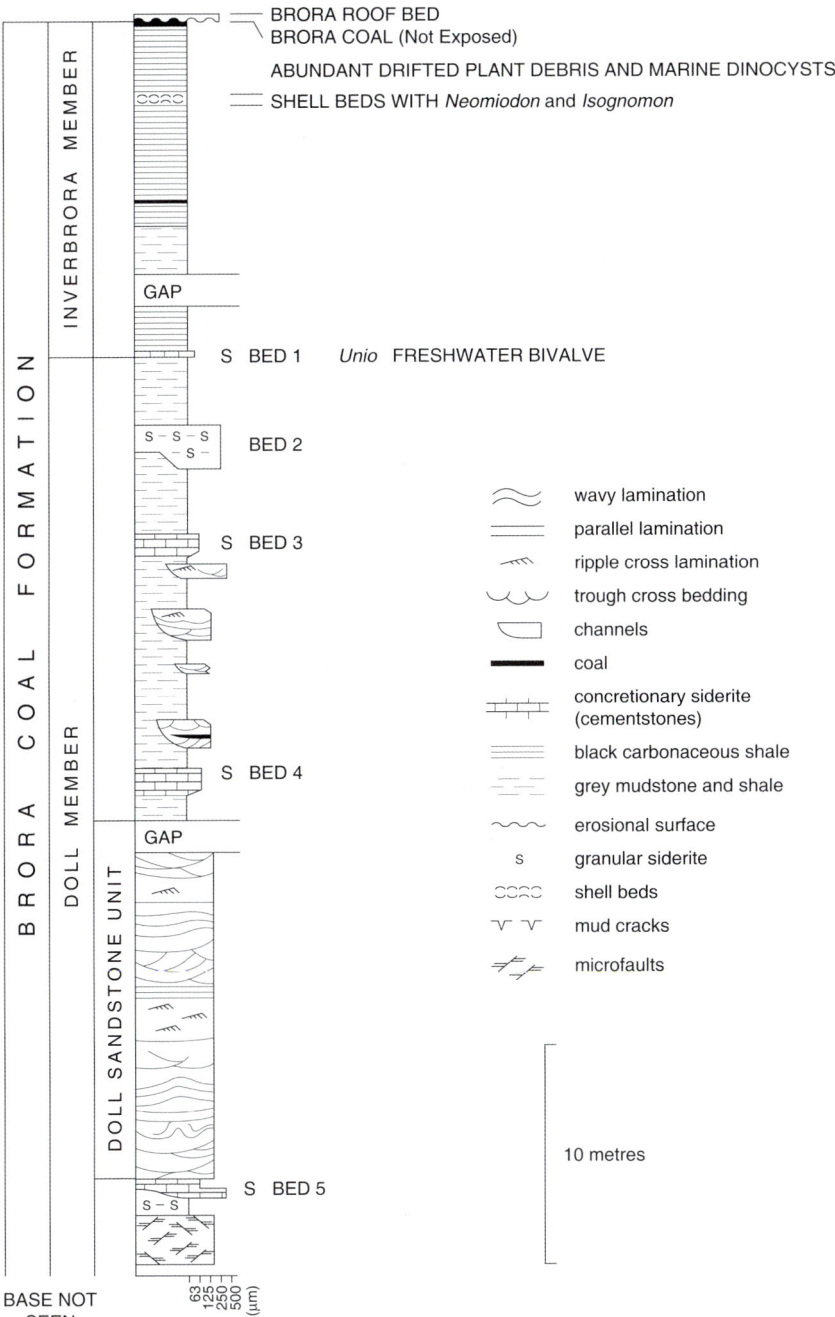

2.5 Sedimentary log of the Brora Coal Formation at locality 1.

lished (Hurst, 1981). The brecciated internal texture of Bed 1 is similar to structures formed by pedogenic (soil-forming) processes, a possibility to some extent confirmed by mineralogical analyses that prove the clay mineral kaolinite is more abundant in the siderite-cemented horizons than elsewhere (Hurst, 1985). The abundance of kaolinite is associated with leaching processes caused by subaerial exposure, often associated with deep weathering. Further confirmation of subaerial processes in the formation of the siderite-cemented beds is the presence of desiccation cracks in Bed 4, and a sandstone bed with rootlets close to Bed 5.

Siderite (iron carbonate) cement is common in freshwater mudstones and provides useful palaeoenvironmental information. As the iron in siderite is present in reduced form it is implied that the cement formed in a reducing environment. Furthermore, the pore water from which the siderite precipitated can be assumed to have been very low in dissolved sulphate, i.e. unlikely to have been of marine origin. In sulphate-rich pore water pyrite (iron sulphide) would form to the exclusion of, or in addition to, siderite. It is interesting to compare the diagenetic mineralogy of the Doll and Inverbrora members. The Doll Member contains diagenetic siderite with no known occurrence of pyrite, whereas the Inverbrora Member contains diagenetic pyrite with no known occurrence of siderite. The mineralogical variation may be attributed to an increased marine influence during deposition of the Inverbrora Member, giving higher sulphate content in the sediment pore waters, an interpretation confirmed by palaeontological data.

The large influx of freshwater ostracods between Beds 3 and 4 (Fig. 2.5) is approximately coincident with a marked change in the clay mineralogy of the mudrocks, from a predominantly kaolinite + illite/smectite assemblage to an illite + kaolinite assemblage (Hurst 1985). No sedimentological evidence is found that corresponds to the palaeontologic/clay mineralogic boundary.

As one continues a stratigraphic descent of the Doll Member it should be noted that the abundance of sandstone increases (Fig. 2.5). At first sandstones occur as small lenses, often less than 5 m wide and only 10 to 30 cm thick. The sandstones are invariably calcite cemented and are easily missed due to the presence of erratics, largely of Devonian age, that litter the shoreline in this area. At the base of the exposed section is the white, fine-grained Doll Sandstone Unit [NC 898 029]. The Doll Sandstone Unit is not calcite cemented; it is lightly consolidated, containing only minor kaolinite and quartz cements, and is estimated to be at least 20 m thick, although the level of exposure often limits observation. All the sandstones are interpreted to be of fluvial origin and contain trough cross bedding with occasional parallel laminae and current ripple lamination (Hurst 1981). The foreset laminae indicate a transport direction from the west or NW with current ripple lamination with more diverse orientation, sometimes normal to the foresets. A derivation of sand detritus from the Barrovian metamorphic rocks of the Grampians to the south or SE was postulated by Hudson (1962, 1964) because of the occurrence of staurolite in the sandstones. In the light of the palaeocurrent data this seems to be unlikely. Possible westerly sources for staurolite are within the Lewisian of the Outer Hebrides (Coward *et al.*, 1969) or from the erosion of post-Barrovian (Devonian) sedimentary rocks deposited west of Brora (Hurst, 1982, 1985).

Close to the line of the Brora Fault and at the south-western extremity of outcrop shown in Figure 2.3, spectacular intensely fractured and micro-faulted outcrop of the Doll Sandstone occurs. The fracturing records the brittle deformation of the Doll Sandstone in the hanging wall of the active Brora Fault. This outcrop is fre-

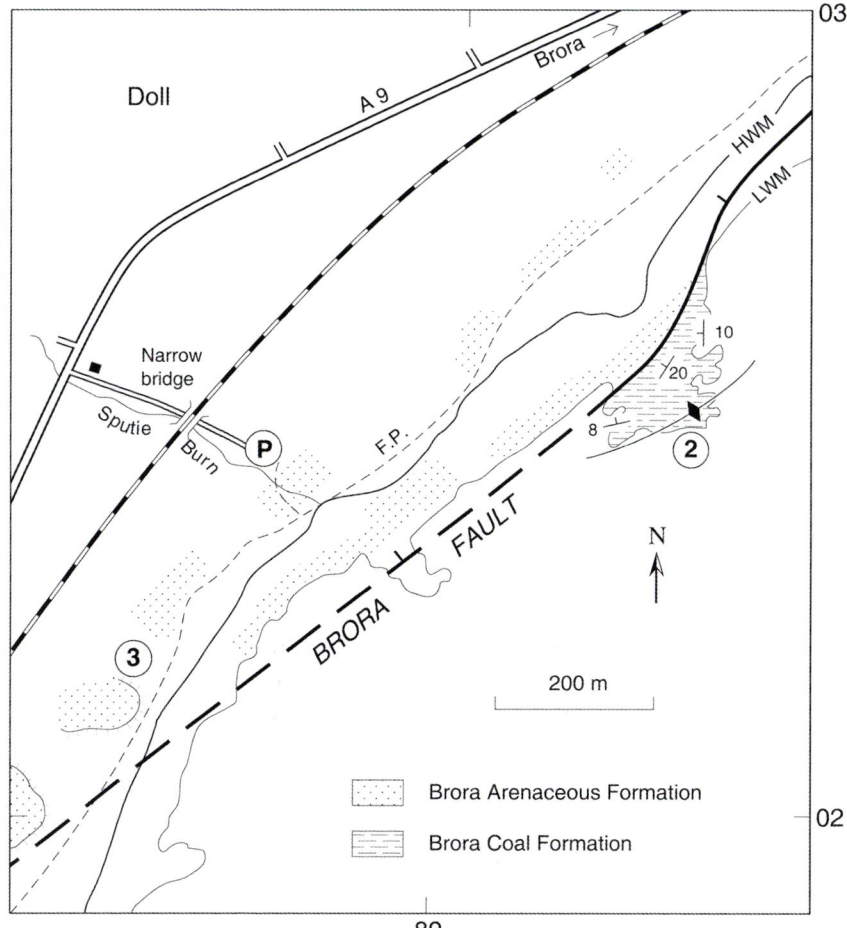

2.6 Sketch-map of the shoreline geology at localities 2 and 3.

quently buried in beach sand. About 80 m east of this area mudstone interbedded with the Doll Sandstone dips at up to 30° to the SE, a local steepening associated with drag against the Brora Fault.

Locality 2. Brora Coal Formation and the Brora Fault near Sputie Burn [NC 893 026]

Continue about 500 m SW to locality 2, which is an intertidal extension of a small headland 500 m NE of the mouth of Sputie Burn (Fig. 2.6). Exposures of the Doll Member are easily accessible at this locality. The lithostratigraphy is as described for the Doll Member at locality 1.

The area of outcrop is the crest of a small anticlinal structure, the axis of which can be observed a few metres above low water mark at the eastern-most tip of the exposure. Unlike the previous section, this section is examined by working up-sequence. The main part of the exposed section dips at angles up to twenty degrees toward WNW–NW. Dips increase toward the NW as the Brora Fault is approached. Outcrop is restricted to the Doll Member, a section from slightly above Bed 1 to the top of the Doll Sandstone Unit usually being visible (Fig. 2.7).

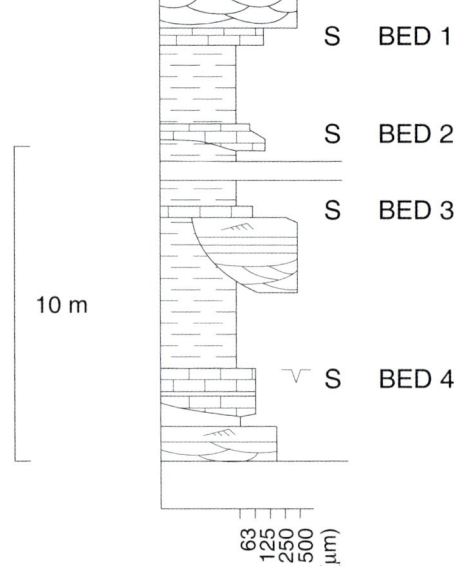

2.7 Log of the Doll Member of the Brora Coal Formation at locality 2 (key as for figure 2.5).

In detail, the sequence is very similar to the main outcrop further NE. A major advantage at this locality is that the exposure is better and less spread than at locality 1. Particularly well exposed are the siderite-cemented mudstones and siltstones that form prominent reefs among the softer mudstones. Examination of the different siderite-cemented beds allows them to be identified as the same 'Beds' shown in Figure 2.4, so allowing lateral correlation to be made. Further confirmation of the stratigraphic position of the section is established by the occurrence of the horizon containing silicified logs between Beds 1 and 2.

Of additional interest at this locality is the exposure of the Brora Fault, which has downthrown the Brora Coal Formation such that it is in contact with sandstones of the Oxfordian Brora Arenaceous Formation. The throw of the Brora Fault can only be approximated, since the precise level of the exposure within the Brora Arenaceous Formation is unknown. However, the entire Brora Argillaceous Formation is faulted out (88.6 m, Sykes, 1975a), the Inverbrora Member is missing (15 m, Hurst, 1981) and the Fascally Member of the Brora Arenaceous Formation (6.5 m Sykes, 1975a) is also missing. The Brora Arenaceous Formation sandstones exposed at Strathsteven (Loc. 3) are believed to be part of the Clynelish Quarry Sandstone Member with the base of the Brora Sandstone Member possibly near the top of the cliffs. The sandstones commonly contain moulds of marine bivalves and some burrows and plant debris, characteristics typical of the Clynelish Quarry Sandstone Member. A throw in excess of 140 m is thus possible. If Lee's (1925) 30 m thickness for the Brora Arenaceous Formation is utilised, a throw of about 100 m is obtained.

Location of the Brora Fault is straightforward as the fault zone is cemented, making the Brora Arenaceous Formation into a more resistant lithology than the surrounding lithologies. Displacement occurred along several sub-parallel planes, and the width of the fault zone and degree of cementation vary. Good examples of oblique slickensides are present on a landward-facing surface by a metal post near high watermark (Fig. 2.8). The fracture is part of the Brora Fault zone. The fault can be traced over most of the exposed foreshore and the fault-associated cementation of the Brora Arenaceous Formation is characteristic. There is no apparent alteration of mudstone lithologies adjacent to the fault zone. Upstanding cemented reefs of Brora Arenaceous Formation occur along the shoreline to the SW beyond Sputie Burn, and mark the continuation of the Brora Fault. The zone affected by

2.8 Exposure of the Brora Fault along the foreshore on the NW edge of locality 2. **A**, the fault is defined by a zone of increased cementation, view of outcrops of quartz-cemented and veined fault rock, and **B**, slickensided surfaces on landward face of outcrop at top of the beach.

cementation seems to vary in breadth from up to 20 m down to apparently nothing toward the NE.

If this itinerary is being completed on foot from Brora, continue SW to the prominent cliff at locality 3.

Locality 3. Brora Arenaceous Formation, Strathsteven Cliff [NC 886 021]

From the car park at Sputie Burn, the locality is some 300 m SW and a path through the disused quarries can be followed (Figs. 2.1, 2.6). Tides do not affect this locality.

At Strathsteven, the thickest continuous vertical outcrop section of the Brora Arenaceous Formation is seen. As no contacts with underlying or overlying formations are visible, it is impossible to know at which stratigraphic level the outcrops occur.

2.9 The Brora Arenaceous Formation at Strathsteven Cliff. Large-scale cross-bedding dipping seawards is seen in what is likely to be part of the Clynelish Quarry Sandstone Member.

Sykes (1975a) assigns the sandstones at Strathsteven to the Brora Sandstone Member, but there are significant differences between these sandstones and those assigned to the Brora Sandstone Member elsewhere in the Brora area (Locs. 6 and 9). The Strathsteven sandstones are generally fine- to medium-grained, well sorted, and rarely contain quartz pebbles. Interbeds of fine-grained sandstone with clay drapes and current ripples are present. Well-developed cross-bedding is uncommon. Moulds of leached bivalves, often brown, are common. These characteristics are more typical of sandstones of the Clynelish Quarry Sandstone Member rather than the Brora Sandstone Member.

Structural measurements are complicated by the lack of distinct bedding planes, a general dip towards SE being inferred. Direction of sediment transport as recorded by cross-bedding has a general trend from NW to SE; however, much variation is present. The NE face of the northernmost cliff has large-scale features that dip to the SE. If these features are sedimentary they can be interpreted as part of a large seaward dipping marine sandbar. The seaward (SE) face of the cliff (Fig. 2.9) exposes a strike section which dips seaward and has elongate lenticular bedding units in which cross-bedding is sometimes identifiable. Bivalve moulds, dominated by pectinids, are common. Good sorting and the removal of any carbonate by leaching have produced high porosity and permeability, averages of 29.8% and 5.1D respectively (based on 18 core plug measurements from the northernmost cliff).

It would be a mistake to concentrate solely on the sedimentological characteristics of the Strathsteven exposures. Indeed, probably their most striking feature is the abundant sub-vertical fracturing. Many fractures sole-out along bedding surfaces; others are vertically continuous at least over the scale of the outcrop. Very few of the fractures have discernible vertical displacements and have probably formed as a result of expansion due to pressure reduction (unloading) during uplift.

At the top of the exposure to the SW [NC 885 020], there is some evidence of fining-upward sequences that are more typical of the Brora Sandstone Member. It is possible that the transition from the Clynelish Quarry Sandstone Member to the Brora Sandstone Member is approximately coincident with the top of the exposures at Strathsteven. The lack of exposure of the Clynelish Quarry Sandstone Member–Brora Sandstone Member boundary elsewhere makes lithostratigraphic division of the Brora Arenaceous Formation problematic. A high degree of lateral and vertical facies variation is expected in marine sand bars; however, it is a mistake to dismiss the problems of correlation because of an inferred depositional complexity.

If the exposures at Strathsteven belong predominantly to the Clynelish Quarry Sandstone Member and the base of the Brora Sandstone Member is near the top of the exposed section [NC 885 020], it would be reasonable to expect that the Fascally Sandstone Member is relatively close in the sub-surface near to the base of the cliffs. In the eventuality of these assumptions being correct, the fine-grained Fascally Sandstone Member and underlying Brora Argillaceous Formation would be effective barriers to surface drainage – a possible reason for the occurrence of the marshy area below and to the south of the cliffs? These assumptions imply that the Clynelish Quarry Sandstone Member is approximately 10 m thick in the Strathsteven area.

As with the exposures of the Clynelish Quarry Sandstone Member and Brora Sandstone Member described at localities 4, 5 and 6, the Strathsteven sandstones are interpreted as marine sand bar facies deposited under the influence of tidal currents. The sandstones were deposited during periods of high depositional energy that were separated by lower-energy periods, possibly involving current reversal, during which clay/silt-rich drapes were deposited. Renewed deposition of sand often eroded both clay drapes and underlying sand units, as recorded by the presence of numerous discontinuous reactivation surfaces.

ITINERARY 2.2
The Brora Arenaceous Formation, south bank of the River Brora

Purpose
To examine exposures of the Brora Arenaceous Formation on the south bank of the River Brora.

Access
Localities 4, 5 and 6 are reached by using public footpaths (Fig. 2.1). Follow the A9 road southward from the centre of Brora until an entrance is reached on the north side of the road beside the house called Catlaw about 50 m beyond Harry Gow's bakery, confectionery and ice cream shop. From this entrance follow a footpath westward through woods overlooking the river until, after climbing over a stile, the path descends to river level at locality 5. Continue upstream following the path high above the river bank; progress may be impeded by vegetation. The path eventually

drops down to the swampy area around Cawcrask at the upstream end of the high bank where exposures are accessible. Locality 4 is about 20 minutes walk from the centre of Brora. Localities 5 and 6 can be reached by walking downstream from locality 4. The limit of tidal influence is at locality 5 but access is unaffected by tides. Locality 6 is affected by tides but is usually accessible at half tide. Locality 6 can also be reached from the centre of Brora by climbing over a partly derelict stile in the wall between the clock tower and Sutherland Arms Hotel. Localities 4, 5, and 6 make a convenient half day excursion.

Introduction

General features of the Brora Arenaceous Formation are described in the geological history section and the basic stratigraphy is illustrated in Figure 2.2. The three members of the formation form an overall coarsening-up sequence onto which are superimposed several smaller-scale fining-up sequences. A marine depositional environment is present throughout.

Sykes (1975a) quotes locality 5 as the type section for the Fascally Sandstone Member. Normally however, none of the Fascally Sandstone Member is exposed on the south side of the river at Fascally. It is assumed that Sykes' true type section is on the north bank of the river opposite locality 5, where a full section of the Fascally Sandstone Member is exposed.

Clynelish Quarry [NC 893 045] (Fig. 2.1), after which the member is named, is not included as part of this guide. The quarry is disused and the site of a scrapyard. Although the rich *lamberti* Zone fauna described by Sykes (1975a) may attract some visitors, do be prepared for a disappointment. The sandstones have tight microcrystalline quartz cement and secondary porosity due to dissolution of *Rhaxella* (sponge) spicules is seen in thin section (Vagle *et al.*, 1994; 1995). Some elements of the fauna can still be collected from small exposures and tip material. A fine ammonite from Clynelish Quarry is placed in the front of the clock tower in the centre of the village. Locality 5 is the type section of the Clynelish Quarry Sandstone Member, and despite not containing a rich fauna it does have excellent exposure of the typical sedimentary facies.

A further exposure of the Clynelish Quarry Sandstone Member, not described in this guide, is at Braamberry Quarry [NC 893 049] (Fig. 2.1). Here, a thicker section is better exposed than at Clynelish Quarry without the tight silica cement. The presence of quartz pebbles and clearly defined cross-bedding are reminiscent of the Brora Sandstone Member, again raising doubts about the internal divisions of the Brora Arenaceous Formation.

Locality 4. Fascally Sandstone Member and Clynelish Quarry Sandstone Member, Cawcrask [NC 899 039]

A gradual increase in dip northwards along the exposure marks the presence of an anticlinal structure between localities 4 and 5. At the southern end of the locality structural dips of less than 10° towards approximately ESE occur.

Sedimentological logs of localities 4 and 5 are given in Figure 2.10. Clearly visible along the entire length of the section is the sharp boundary between the

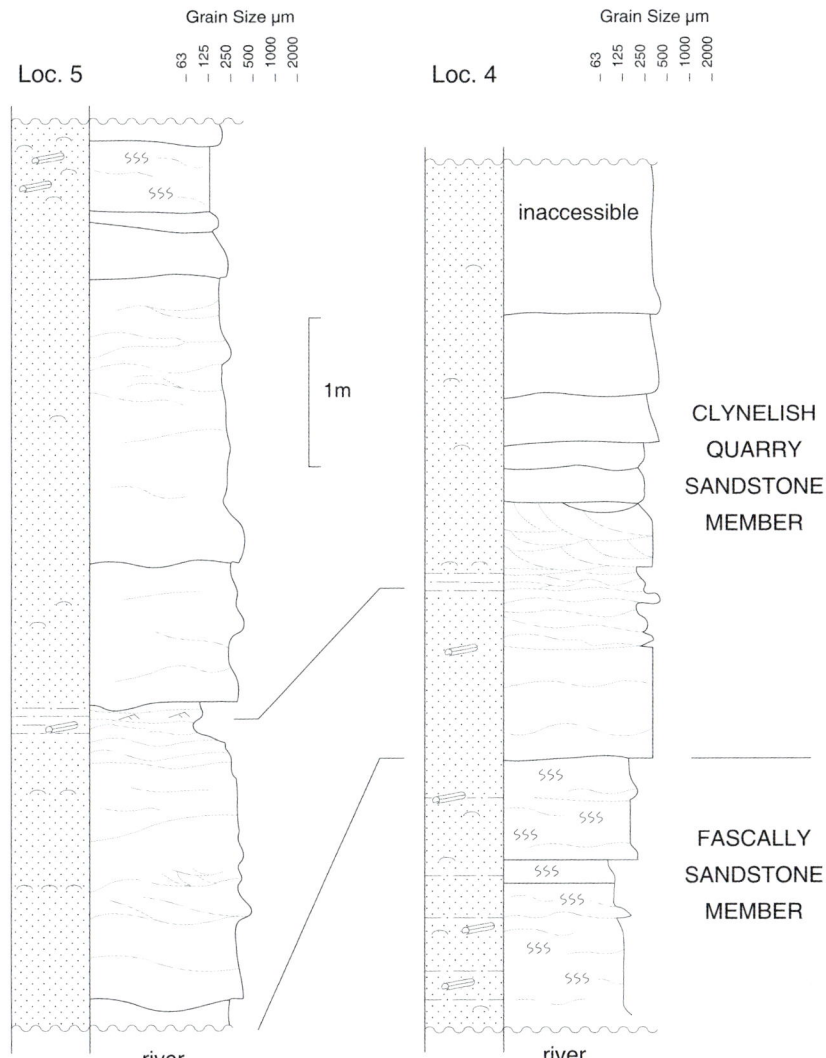

2.10 Sedimentary logs of the Clynelish Quarry Sandstone Member and Fascally Sandstone Member at localities 4 and 5. (Key as for figure 2.2.)

Fascally Sandstone Member and the Clynelish Quarry Sandstone Member. The Fascally Sandstone Member is a fine- to very fine-grained silty sandstone of marine origin. Belemnites and other shelly marine fauna, especially the bivalve *Chlamys*, are common, although the original shell material has been removed by leaching, and fossils occur as iron-oxide stained moulds. Few primary structures are visible, as the Fascally Sandstone Member is intensively bioturbated, networks of sub-horizontal burrows predominating. Sykes (1975b) notes the presence of *Thalassinoides*. Traces of wave-ripple lamination are distinguishable. On two core plugs porosities and permeabilities of 24.7% and 28.7% and 93 mD and 27 mD were obtained. Better exposure of the Fascally Sandstone Member occurs on the north bank at locality 5 and can be visited in association with Itinerary 3 of this excursion.

The increase in grain size between the Fascally Sandstone Member and Clynelish Quarry Sandstone Member is small, from fine to medium sand <125 μm to <250 μm (Fig. 2.10, Loc. 4). Most obvious is the decrease in detrital clay content, exemplified by the colour change from the grey-green Fascally Sandstone to the pale yellow Clynelish Quarry Sandstone, which is well sorted, fine-grained and highly quartzose. At this locality traces of bioturbation are rare and fossils are restricted to leached fragments of bivalve shells. In equivalent strata on the foreshore immediately south of the river, bioturbation is well preserved, including large (approximately 20 cm diameter) vertical burrows. Preservation of fossils and bioturbation seems to be associated with the presence of an early diagenetic quartz cement.

A very slight overall coarsening-upwards sequence is observed in the Clynelish Quarry Sandstone Member, although individual bedding units often contain fining-upward sequences. At first sight the sandstones appear to be massive; however, some intervals are cross bedded and indicate a transport direction from WNW/NW. No channel form is recognised and each sandstone unit rarely exceeds 1 m thickness. The apparent lack of sedimentary structures is partly because the outcrop is orientated parallel to the depositional trend, which makes them difficult to see, and partly because the sandstones are very well sorted.

Detrital and authigenic clays are uncommon in the Clynelish Quarry Sandstone Member sandstones, which are classified as quartz arenites (Hurst, 1980). Clay-rich laminae are, however, common along the bedding surfaces between sandstone units. Frequently, the laminae are little more than 1–2 mm thick and contain finely-disseminated plant remains. The plant remains may previously have been pyritised, as they now have a rusty coloration caused by weathering. Approximately 1 m above the base of the Clynelish Quarry Sandstone Member is an interval with a concentration of muddy laminae. This interval is about 25–30 cm thick; however, thickness and internal characteristics vary laterally. In general, the interval comprises very fine-grained sandstones within which muddy lenses of fine- to medium-grained sandstone occur (Fig. 2.10). The muddy layers are undulose and drape onto sandstone units. Some sandstone units are apparently enveloped by muddy laminae. Fine-grained sandstones within the muddy laminae sometimes contain current ripples. The muddy laminae are sometimes disrupted laterally by down-cutting erosive surfaces filled with sandstone.

Sykes (1975a) makes no specific sedimentological interpretation of the Clynelish Quarry Sandstone Member, the whole of the Brora Arenaceous Formation being interpreted as a coastal sandbar. The action of tidal processes can, however, be inferred with confidence based on the occurrence of mud drapes, the sharp erosive base of the member, the dominance of unidirectional current structures and the marine environment. Evidence of current reversal is not readily observable at this locality, as current ripples in the muddy intervals are too few and too indistinct to make a conclusive interpretation. No evidence is preserved of wave-influenced processes. A sub-tidal marine depositional environment is inferred with no direct evidence for a coastal affinity.

Locality 5. Clynelish Quarry Sandstone Member, Fascally [NC 899 040]

This locality lies on the northern limb of a gentle SE plunging anticline. Dips of around 15° toward SE are probable. An outcrop of the Fascally Sandstone Member on the opposing bank of the river illustrates the structural dip more clearly than in the Clynelish Quarry Sandstone Member.

The sequence is similar to that exposed at Cawcrask (Loc. 4) as comparison of the sedimentary logs shows (Fig. 2.10). A horizon of very fine-grained sandstone with muddy laminae is clearly visible about 2 m above the river level. This horizon may correlate laterally with similar laminated sandstones observed 1 m above the contact with the Fascally Sandstone Member at locality 4. Within the fine-grained sandstone interval current ripples are observed that appear to have formed in currents moving from ENE to WSW, i.e. slightly oblique to perpendicular in relation to the direction of the foresets in the cross-bedded sandstones. Structures within this laminated interval provide evidence of current reversal. Cross-bedded intervals are rarely clearly visible at this locality; however, the same uneven bedding thicknesses as seen at locality 4 are observed. Concentrations of leached shell debris are locally common.

Average porosity and permeability of 30.1% and 8.2 D respectively are measured on core plugs from locality 5 (14 samples). Permeability increases upward in the section.

Above the laminar sandstone the vertical sequence is examined by moving southward toward the path. The remaining sequence shows an overall fining-upward trend in which bedding surfaces are often discontinuous. Muddy layers and mud drapes are most common near the top of the sequence, whereas shell debris is more common in the lower part of the sequence. Near the top of the exposed section, along the upper side of the path, approximately 0.5 m of very fine-grained bioturbated sandstone, similar in appearance to the Fascally Sandstone, is found. Interbedding of bioturbated facies within the otherwise clean Clynelish Quarry Sandstone Member was not recorded by Sykes (1975a, b).

The lower part of the Clynelish Quarry Sandstone Member and the top of the Fascally Sandstone Member are also exposed on the north side of the river opposite the harbour [NC 908 040 to 910 039], where thin yellow sandstone beds and interbeds of muddy bioturbated sandstones are exposed. This locality is best visited at low tide.

Locality 6. Brora Sandstone Member, Brora [NC 906 039]

Continue downstream to locality 6, which is immediately upstream of the road bridge. Approximately 5 m of the Brora Sandstone Member is exposed at locality 6. If the tide is low, riverside exposures can be examined between localities 5 and 6. The rock by the river is very slippery and extra care is needed. A large cliff exposure, which is only fully accessible at low tide, is located about 200 m upstream from locality 6 [NC 903 039]. At locality 6 a very low structural dip is apparent, the bedding surfaces along the river bank sloping at an angle similar to the surface of the river (<1°). It is of course very difficult to measure structural dip in sandstones that contain sedimentary structures that produce a bedding structure.

At the cliff face and along the river toward the bridge the Brora Sandstone Member comprises predominantly planar cross-bedded pebbly quartzose sandstones. From the terrace exposed below high-water mark up to the cliff face there is an overall coarsening upward. Bed thicknesses vary laterally and can be traced downstream along the riverside at low tide. Outcrops at the same stratigraphic level within the Brora Sandstone Member are found on the downstream side of the bridges and may be approached from the harbour. The cliffs on the north side of the river are not accessible but display similar features to outcrops at locality 6.

Unlike locality 5 no muddy horizons are seen and planar cross-bedding predominates. The approximate current direction derived from the orientation of the cross beds is from ENE, the opposite direction to that measured in the Clynelish Quarry Sandstone at Fascally (Loc. 5). Leached shell fragments occur sporadically and sometimes form coarse lags at the base of foresets. Well-rounded quartz pebbles are characteristic both as lags and as isolated grains within otherwise poorly sorted medium- to coarse-grained sandstone. Sedimentary bedding characteristics are sometimes confused by the presence of low-angle fractures, which often sole out along bedding planes.

Average porosity and permeability at locality 6 are 31.3% and 4.6 D, respectively. Large ranges of porosity and permeability are present, 27% to 37% and 2.4 D to 12.9 D, respectively.

The Brora Sandstone Member is interpreted as being part of a tidal sandbar complex, probably a facies type that is part of both a lateral and, more often, a vertical facies association with the Clynelish Quarry Sandstone Member. The absence of mud drapes, and fine-grained material in general, is evidence of the continuous high-energy depositional environment which presumably removed any mud drapes deposited under still-stand conditions.

ITINERARY 2.3

The Brora Argillaceous Formation, north bank of the River Brora

Purpose

To examine the exposures of the Brora Argillaceous Formation exposed on the north bank of the River Brora.

Access

From the north side of the A9 road bridge in the centre of Brora take the road westwards signed Gordonbush and Rogart (Fig. 2.1). After slightly less than 1 km the recreation area and car park is reached. This is a sensible place to park although it is possible to drive a car up to Bruachrobie. From the car park, either the track to the east or the west of the pitch can be followed to Bruachrobie (views of locality 4, Brora Arenaceous Formation on the south bank of the river from the eastern path – remains of old brick clay pits from the western path). The locality is most readily accessible from the western end [NC 8875 0398], which is reached by following the overgrown track past Bruachrobie to the top of hill following the boundary fence approximately westward through several gates until a marked break of slope is reached. River level can be reached via a small gate near the edge of the cliff, giving access to outcrops of the Brora Brick Clay Member. Wellington boots or waders are recommended for examination of this locality, which is only accessible when the river is low.

Approximately 3 hours should be allowed to complete this itinerary.

Introduction

The Brora Argillaceous Formation is nowhere well exposed. The shore section (Itinerary 1) is only visible at low tide, complicated by faulting and frequently obscured by seaweed and beach sand. Inland exposures have weathered surfaces that obscure many sedimentary features. Sykes (1975a, b) described the shore sections and sections from Coal Board boreholes. Landscaping in the Fascally area [NC 898 041] has destroyed sections once available in the Brora Brick Works pit described by Sykes (1975a, b). Biostratigraphic data are found in Sykes (1975a) and Lam and Porter (1977). Some details of the Brora Shale Member on the shore are given by MacLennan and Trewin (1989), and in Itinerary 1 of this excursion.

Locality 7. Brora Argillaceous Formation, near Bruachrobie [NC 888 040]
Brora Shale Member

About 80 m downstream from the western end of the cliff the top of the Brora Shale Member is exposed.

Approximately 2 m vertical thickness is exposed before thick vegetation obscures further examination. Exposure of the Brora Shale Member forms the core of an anticlinal structure the eastern limb of which takes exposures of the Glauconitic Sandstone and Brora Brick Clay members down to river level further downstream. Plant fragments, belemnites and phosphatic nodules are common in the sandy horizons. The sandy beds are interpreted to be the deposits of gravity-flows. They contain glauconite and a shallow-marine shelf bivalve fauna, presumably reworked by tidal or storm processes, and redeposited in a more stagnant deeper-water environment. Sykes (1975b) records the presence of the trace fossils *Thalassinoides*, *Diplocraterion* and *Chondrites* from the sandy beds, but they are absent from the shaly interbeds. Clearly, the background anoxic environment in which the shales were deposited was unsuitable for sustaining infaunal activity.

Glauconitic Sandstone Member

The top of the Brora Shale Member is marked by a rapid coarsening-upwards into thick (0.5 m) glauconitic, very fine-grained sandstone (Fig. 2.11), which marks the base of the Glauconitic Sandstone Member (Sykes 1975a). Three subdivisions are defined within the Glauconitic Sandstone Member: a basal unit comprising $c.4.15$ m of muddy glauconitic sandstones with thin shale interbeds; a middle unit comprising $c.2$ m of grey siltstone; an upper unit comprising $c.3.9$ m of slightly glauconitic silty sandstone. The siltstones have a sparse fauna but the sandstones contain abundant fossils and are bioturbated.

In the lowermost 4.15 m of the Glauconitic Sandstone Member eight small coarsening-upward sequences are identified. The sandstones are similar and characterised

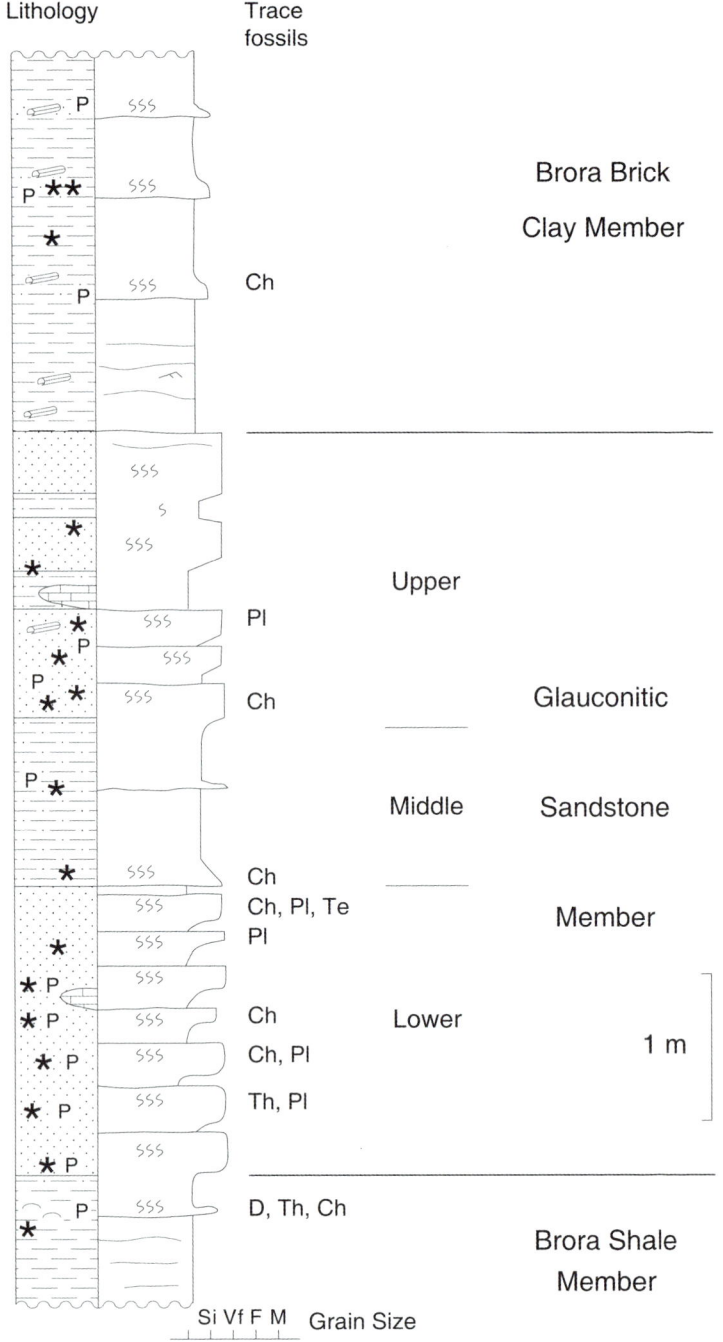

2.11 Sedimentary log of the Brora Argillaceous Formation at locality 7. Key as for 2.2. Pl, *Planolites*; Ch, *Chondrites*; Te, *Teichichnus*; Th, *Thalassinoides*; D, *Diplocraterion*.

by rows of phosphatic nodules (cm-scale) and horizons of randomly-orientated tree fragments and guards of the belemnite *Cylindroteuthis puzosiana* (Fig. 2.12). The phosphatic nodules contain sponge spicules of *Rhaxella perforata*, which are visible

2.12 Belemnites (*Cylindroteuthis*) that are common in the Glauconitic Sandstone Member and adjacent strata.

in thin section (Sykes 1975a). Despite the pervasive, intense bioturbation, the trace fossils *Thalassinoides*, *Diplocraterion* and *Chondrites* are identified at sandstone–shale boundaries (Sykes, 1975b).

The sandstones are silty with a high content of sand-sized glauconite grains. The glauconite is allochthonous, having formed at a sediment–water interface in gently agitated water that periodically disturbed poorly-oxygenated bottom waters. Faecal pellets and small shell fragments have probably been pseudomorphed by the glauconite. The glauconite has similar smectite content (30–40%) to clays in the surrounding siltstones and shales (Hurst, 1985). This similarity is indicative of the compositional uniformity of the source of clay detritus to the Moray Firth Basin during Callovian times. Deposition of the sandstones is interpreted as being the result of progradational events higher up on the shelf that caused the deposition of sand further offshore, perhaps during storm periods. Deposition of bituminous silty mud with a sparse infauna is indicative of poor circulation and poor oxygenation at the sediment–water interface. Sand deposition records periods of increased depositional energy and increased oxygenation when burrowing organisms colonised the area until anoxic bottom conditions were re-established. Mud deposition appears to have excluded the survival of an infauna in this lower unit.

Grey siltstones in the middle unit of the Glauconitic Sandstone Member become more silt-rich upward, the lowermost beds being dark and apparently bituminous. Fauna is scarce and restricted; Sykes (1975b) recorded the presence of the bivalve *Meleagrinella braamburiensis* and *Lingula craniae*. It is assumed that a break occurred in the progradation represented by deposition of the underlying unit, thus re-establishing fine-grained deposition. The decreased bituminous content upward is interpreted to represent the onset of more oxic conditions at the sediment–water interface.

Sandstones in the upper 3.9 m of the Glauconitic Sandstone Member are less glauconitic than in the lower unit, thus giving a cleaner appearance. Five coarsening-upwards beds are identified (Fig. 2.11), the top bed being 0.8 m thick. Prominent carbonate concretions ('doggers') occur about 2 m from the top of the Glauconitic Sandstone Member. Sykes (1975a) suggests that the upper unit of cleaner sandstones in the Glauconitic Sandstone Member is the product of a period of reworking, evidence for which can be seen at Balintore in Ross-shire. Certainly the lower glauconite content of the sandstones is indicative of better oxygenated conditions in the shallower part of the basin from where it was derived. The shaly interbeds have more evidence of bioturbation than is seen lower in the sequence, again indicative of oxic conditions.

Brora Brick Clay Member

Approximately 5 m of an estimated total thickness of 15 m of the Brora Brick Clay Member crop out along the riverbank. The Brora Brick Clay Member is the finest-grained member of the Brora Argillaceous Formation and was until 1974 extracted in the Fascally area for making bricks.

The base of the Brora Brick Clay Member is marked by the abrupt change in grain size at the top of the last sandstone bed of the Glauconitic Sandstone Member. Laminated, slightly bioturbated muddy siltstones comprise the basal 1.7 m of the Brora Brick Clay Member. Fauna is sparse, Sykes (1975b) recording the occurrence of belemnites (*Cylindroteuthis puzosiana*) and the bivalve *Solemya woodwardiana*. Three thin, silty, very fine-grained sandstone beds may be visible near the top of the exposed sequence, which are the bases of three fining-upward units. Exposure of the upper part of this sequence is poor due to slippage and is often inaccessible if the level of the river is high.

The thin fining-upward sequences in the Brora Brick Clay Member are interpreted as being the result of deposition from small gravity flows. They are similar to sandstones in the Glauconitic Sandstone Member and contain an allochthonous fauna and glauconite derived from a nearer-shore environment.

Locality 8. Brora Brick Clay Member, west side of recreation area [NC 896040]

Access to this clay pit (Fig. 2.1) is now closed. It is the only clay pit of the brickworks that survived the landscaping of the Fascally area. The exposures have uneven weathered surfaces.

Locality 5. Fascally Sandstone Member [NC 899 040]

Depending on the river level access can be made on the north side of the river (Fig. 2.1) either from the downstream side, to examine the Fascally Sandstone Member, or from the upstream side via the concrete embankment, to examine the top of the Fascally Siltstone Member. The locality is only accessible when the river is low.

A prominent series of calcareous nodules form a useful marker horizon that is approximately 1 m from the top of the Fascally Siltstone Member. The boundary between the members is marked by the first occurrence of intensely bioturbated, fine-grained, muddy sandstones (Sykes 1975a). The features of the sandstones have been described previously (Loc. 4) and are not repeated here.

ITINERARY 2.4
The Brora Arenaceous and Balintore formations, north foreshore

Purpose
To examine the Brora Sandstone Member at the top of the Brora Arenaceous Formation and the Ardassie Limestone Member of the Balintore Formation.

Access
Take Golf Road eastward from the north side of the bridge in the centre of Brora to the golf club car park. A public footpath crosses the golf course to the beach. The rocks at locality 9 are covered at high tide, but some exposure remains at mid-tide. Sand cover severely restricts the level of exposure of the Brora Sandstone Member. The itinerary takes 1 hour or more depending on how much time is spent looking for fossils. Please do not hammer the outcrops; material can usually be found loose at the top of the beach at Ardassie Point.

Introduction
On the northern foreshore at Brora exposures lie between the tide marks (Fig. 2.13). Approaching Ardassie Point from the south as much as 10 m of the Brora Sandstone Member may be exposed together with approximately 12 m of the Ardassie Limestone Member (Sykes 1975a).

Locality 9. Brora Sandstone Member and Ardassie Limestone Member, Ardassie Point [NC 913 041]
Despite the general overall upward-coarsening of the Brora Arenaceous Formation (Fig. 2.2) this section has a slight overall fining-upwards trend that continues into the fine-grained sandstones and calcareous mudstones of the Ardassie Limestone Member. The lowest 2–3 m of the exposed section (often poorly exposed) comprises medium-grained sandstones with occasional pebbly or shelly bottom-sets but lacking any well-defined lamination. Bedding planes are irregular and elongate dune forms are identified. Throughout the remaining part of the section trough cross bedding predominates with individual trough forms identifiable. Palaeocurrent directions are toward the W and SW. Rounded quartz pebbles and phosphatic fragments are found in the lags of some foresets.

Smaller scale fining-upward sequences are identified within the overall fining-upward sequence. Near the top of the Brora Sandstone some vertical burrows are present, which probably record the initiation of deeper-water conditions.

The boundary between the Ardassie Limestone Member and the Brora Sandstone Member is distinct, changing from clean yellow sandstones to grey, very fine-

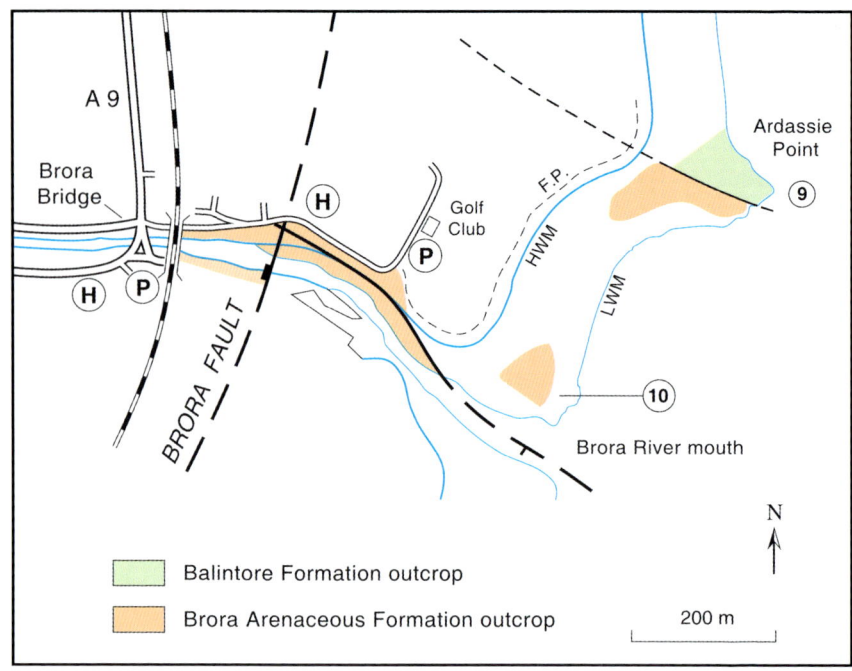

2.13 Locality map for localities 9 and 10, Ardassie Point and estuary of the Brora River.

grained, shaly sandstones. The contact is rarely exposed. Sykes (1975a) describes the rocks at Ardassie Point as muddy carbonaceous sandstones interbedded with slightly sandy spiculites, originally with opaline sponge spicules (*Rhaxella*), which are now replaced by calcite. The sandstones are thoroughly bioturbated and fossiliferous, large specimens of the bivalves *Pinna lanceolata* and *Gryphaea dilatata* occur preserved in life position, and the bivalve *Cucullaea* is common. Sykes (1975b) records the presence of a rich ammonite fauna typical of the *vertebrale* Subzone, which includes *Cardioceras (Subvertebriceras) densiplicatum, C. (Subv.) sowerbyi, C. (Scoticardioceras) excavatum* and *C. (Plasmatoceras) tenuistriatum*. Faunal diversity and content decreases upward. Sykes (1975b) notes that the less diverse fauna at the top of the sequence is dominated by the bivalve *Chlamys (Radulopecten) fibrosa*.

Sedimentologically the sandstones of the Ardassie Limestone Member resemble the Fascally Sandstone Member with their characteristic mud-filled burrows and occasional traces of wave-generated structures. Because of the tight cementation in the 'limestones' it is impossible to measure any grain-size variations, but they are finer grained than the sandstones and may represent the tops of a series of fining-upwards units that occur throughout the Ardassie Limestone Member. The resumption of fine-grained sedimentation may be a response to a deepening of the basin and/or restriction of sand supply.

Deposition of the Ardassie Limestone Member is interpreted to have taken place on a marine shelf which was undergoing a period of regional transgression. Sandstones are thought to have been deposited either during storm periods or, during spring tides, winnowed by waves and finally colonised by organisms.

Locality 10. Faulted sandstones [NC 911 038]

From locality 9 return southward toward the river estuary (Fig. 2.13). At low tide exposures of sandstones of the Brora Arenaceous Formation can be examined on the foreshore. The outcrops are resistant due to quartz cementation and the presence of cemented, slickensided fractures. These features are evidence of a zone of faulting oriented between SE–NW and E–W.

The fault zone follows the line of the river and can be traced along the north riverbank and through the dilapidated picnic park in front of the Royal Marine Hotel. Quartz cementation associated with the fault zone is also present on the south side of the river mouth.

Excursion 3

The Upper Jurassic of the Helmsdale area

A. C. MacDonald and N. H. Trewin

Purpose

To illustrate sections across the Helmsdale Fault zone and examine the boulder beds and other Upper Jurassic lithologies deposited in response to fault movement at the western boundary of the Inner Moray Firth Basin. The excursion is divided into the following four itineraries.

Itinerary 3.1 Kintradwell. The Kintradwell Boulder Beds with slump and slide features and an intruded sandstone dyke.

Itinerary 3.2 Lothbeg Point area. The Allt na Cuile Sandstone, and Lothbeg Siltstone. Sand from a delta transported to deep water.

Itinerary 3.3 Portgower. Boulder beds with giant clasts; the 'fallen stack' locality, and changes in clast types within the boulder beds.

Itinerary 3.4 Helmsdale. The Helmsdale Boulder Beds, rockfall breccias, and Helmsdale Fault outcrop.

General access

Details of access are given with each of the itineraries but the following notes are generally applicable. The itineraries cover most of the major features seen in exposures between Kintradwell and the Ord of Caithness. The whole excursion requires about 2 days and is best carried out in the spring or early summer when vegetation is at a minimum. The shore sections are tide dependent, thus it is best to plan the trip when low tide occurs during midday and visit the localites in an appropriate order. The use of a car or minibus will allow fairly easy access to all localities. If a coach is used the party will have to be dropped at the main road and a little more walking will be needed. The sequences crop out along the rocky foreshore, in river gullies and on raised beaches. Where appropriate, permission should be sought from farmers or landowners to cross agricultural land. Two short itineraries from the first edition of this guide have been omitted due to safety and access problems. Do not attempt to visit exposures in rail cuttings (illegal and dangerous), and take great care crossing the rail line at the designated crossings. Trains are infrequent, but surprisingly silent, and many are not timetabled. The locations of the areas covered by the itineraries are shown on Figure 3.1 and the main stratigraphic features in Figure 3.2.

3.1 *Opposite, above:* Location areas of itineraries 1 to 4 of Excursion 3.
3.2 *Opposite, below:* Basic stratigraphy of the Kimmeridgian section with approximate stratigraphic positions of localities described in the excursion.

Excursion 3
Location areas of Itineraries 1-4
A 897
R. Helmsdale
A 9
HELMSDALE
4
15
Portgower
N
3
A 9
1 KINTRADWELL
2 LOTHBEG
3 PORTGOWER
4 HELMSDALE
10
Lothbeg Point
2
Kintradwell
0
5 Km
A 9
1
95
00
05

STAGES
ZONE
PORT.
KIMMERIDGIAN
LATE
EARLY
VOLGIAN
KIMMERIDGIAN
MIDDLE
LOWER
albani
fittoni rotunda
pallasioides
pectinatus
hudlestoni
wheatleyensis
scitulus
elegans
autissiodorensis
eudoxus
mutabilis
cymodoce
baylei
Itin 4
Helmsdale Boulder Beds c. 800 m
Itin. 3
?
Itin. 2 Allt na Cuile Sandstone c. 120 m
Itin. 1 Kintradwell Boulder Beds c. 85 m
Dominant ORS flagstone clasts
Dominant Sst. clasts
Rockfall breccias adjacent to Helmsdale Fault
Boulder Beds (diagrammatic)
Channel to parallel bedded facies, Allt na Cuile Sst.

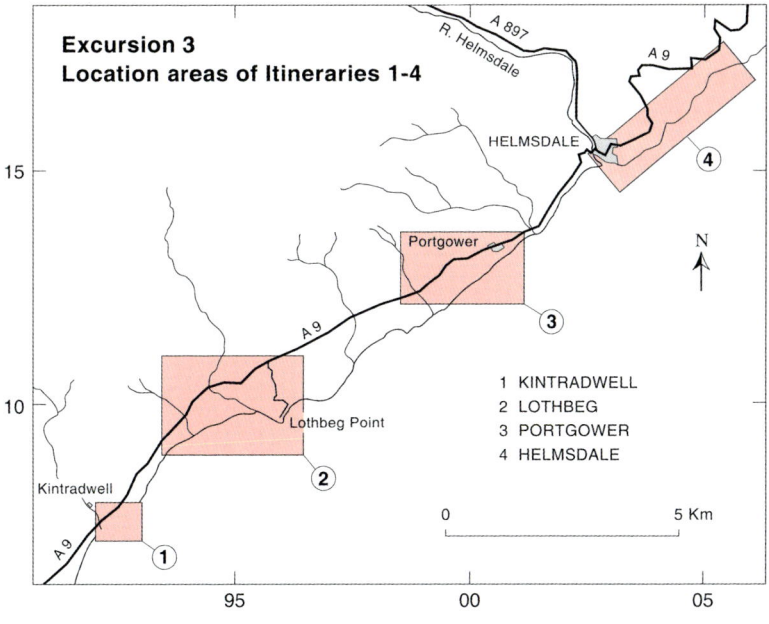

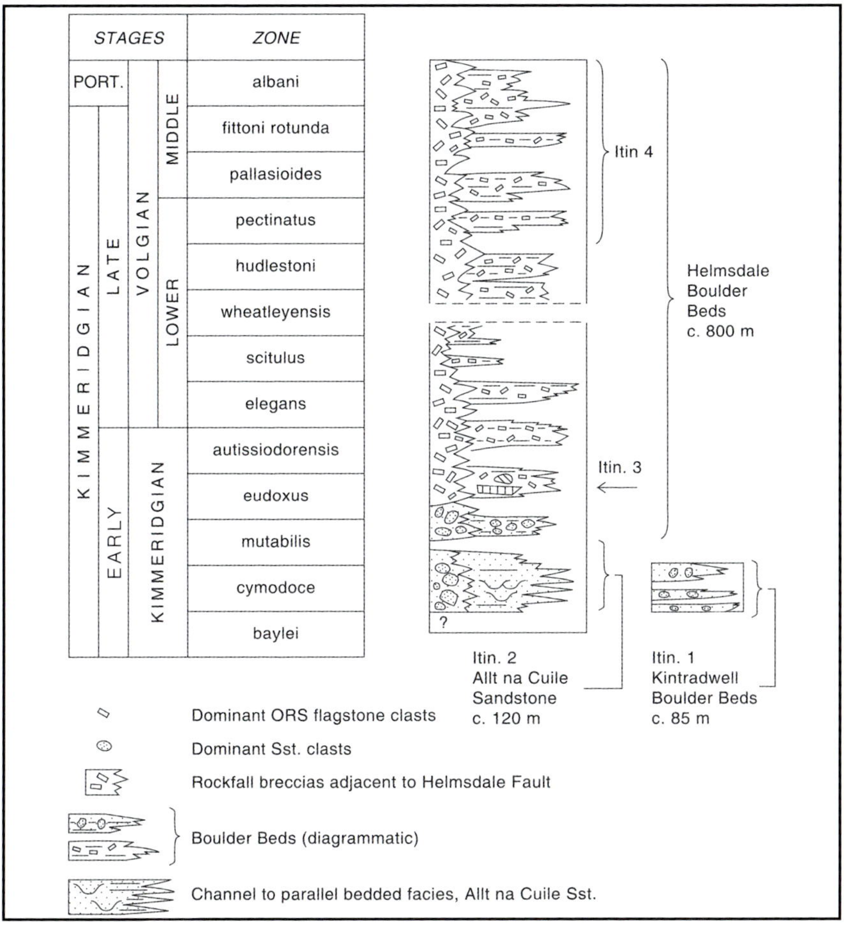

General introduction

The sequences are exposed in a narrow coastal strip up to 1.5 km in width between Kintradwell and the Ord of Caithness. The sediments are downfaulted against Moinian granulites, Helmsdale Granite and Old Red Sandstone to the west, and are bounded by the Moray Firth to the east. There is an overall younging of the sequence from SW to NE, from Kimmeridgian *cymodoce* Zone at Kintradwell to Middle Volgian *albani* Zone near the Ord of Caithness (Lam and Porter, 1977; Riley, 1980).

The late Jurassic of the Inner Moray Firth was characterised by rapid subsidence and the formation of half-grabens (Fig. 8 in Geological History section). The sediments exposed in the Helmsdale area were deposited at the western active margin of one of the major half-grabens controlled by downthrow on the Helmsdale Fault. These excursions illustrate the spatial and temporal distribution of lithologies in this fault-bounded basin-margin setting and highlight some of the unusual sedimentary features which developed on the fault-controlled submarine slope.

It is difficult to measure accurately the thickness of the sequence because of discontinuous outcrops, folding, small-scale faulting and disconformities. The Kimmeridgian to Middle Volgian sequences probably total more than 900 m (data in Wignall and Pickering, 1993). In well 11/30–1 of the nearby Beatrice Oilfield the basal Kimmeridgian to late Volgian sequence is approximately 750 m thick and dominantly of shale. The accumulation of these thick sequences illustrates the rapid fault-controlled basin subsidence during the late Jurassic.

From the Lias through to the Oxfordian the Jurassic sequences of East Sutherland are of shallow marine to freshwater origin. Large-scale differential movements on the Helmsdale Fault began during the early Kimmeridgian and led to the establishment of a deep-water submarine slope environment on the downthrown side to the SE of the fault, and a relatively stable shallow-water platform area on the upthrown side to the NW. A spectacular submarine fault-scarp developed between the shallow and deep-water areas. Debris from the fault-scarp formed a submarine talus slope, and sand, shell fragments and land-derived plant debris were washed over the fault from the shallow-water shelf environment into the deep-water environment where suspension settling of mud was the background sedimentation. At the foot of the talus slope, formed largely of rockfall breccias, accumulation took place of boulder beds emplaced by debris flows and other gravity-flow mechanisms. Initiation of gravity flows was probably triggered by earthquakes caused by movements on the Helmsdale Fault. It is thus reasonable to hypothesise that tsunamis would have resulted from sea-floor movements and the narrow shallow shelf bordering the Scottish Landmass would have been severely affected. The great mix of material in the individual beds, including both marine organisms and land-derived plants, would support initiation of debris flows following tsunamis. This scenario has been woven into a time-travel excursion by Trewin (2008) which may be of interest to promote further debate on the origins of the boulder beds.

The boulder beds are interbedded with siltstones, mudstones and thin sandstones. The sand was probably swept off the adjacent shelf by storm, wave and tidal action. The thin sand beds and laminae in the shales between the boulder beds

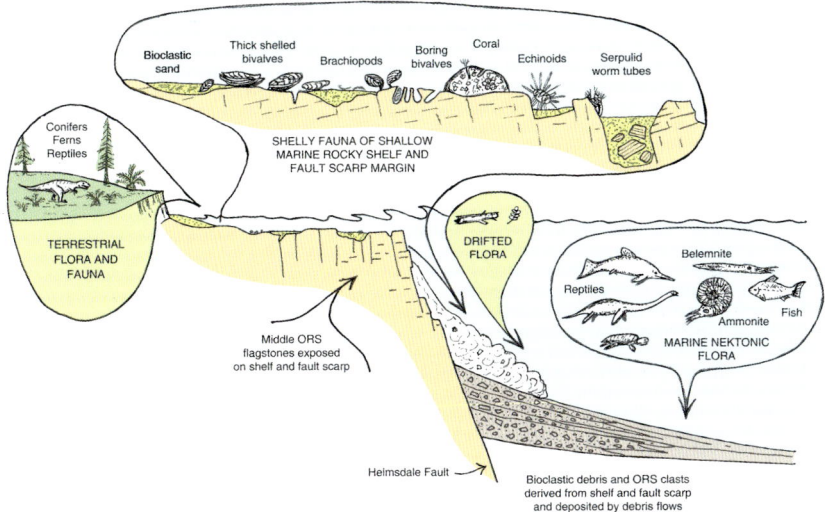

3.3 Cartoon showing origin of fauna and flora associated with the Helmsdale Boulder Beds.

greatly exceed the boulder beds numerically, and were clearly the result of frequent events such as storms.

The displaced fossils in the deep-water deposits include a variety of shallow marine bivalves, brachiopods, echinoid debris and corals. Land-derived debris in the form of wood and leaf material (Seward, 1911; Van der Burgh and Van Konijnenburg-Van Cittert 1984; Van der Burgh 1987) became waterlogged and sank into deep water to be preserved in shales along with an open marine fauna of ammonites and belemnites (Fig. 3.3). Bottom conditions on the downthrow side of the fault were reasonably well oxygenated during deposition of the Kintradwell Boulder Beds and a diverse bivalve fauna is present (Wignall and Pickering, 1993). Later in the Kimmeridgian oxygenation was poorer and the benthic fauna is greatly reduced.

Clasts within the Boulder Beds reflect the strata that were exposed on the active fault scarp at the time of deposition. The youngest clasts, possibly reworked from the Jurassic, together with sandstones, probably from the Upper ORS, occur in the oldest (*cymodoce* Zone) boulder beds in the south and Middle ORS flagstone clasts in the youngest boulder beds at the north of the outcrop.

Submarine fans were not extensively developed along the fault-margin. Sand-rich sequences are limited to *cymodoce* Zone and were deposited contemporaneously with the initiation of the large-scale faulting episode. These complex sequences may represent a localised submarine canyon and fan body which was fed from a small delta on the adjacent shelf (Fig. 3.4). Generally, the development of submarine fans was precluded by the lack of coarse-grained sediment supply to the high-energy rocky shelf area which developed to the NW of the fault scarp. The term 'submarine fan' is used here in the most general sense. It is not possible to define the shape of the sand body, or to say whether it was multi-sourced from the fault scarp, or channelled through one or more point sources. It seems probable that sand supply was control-

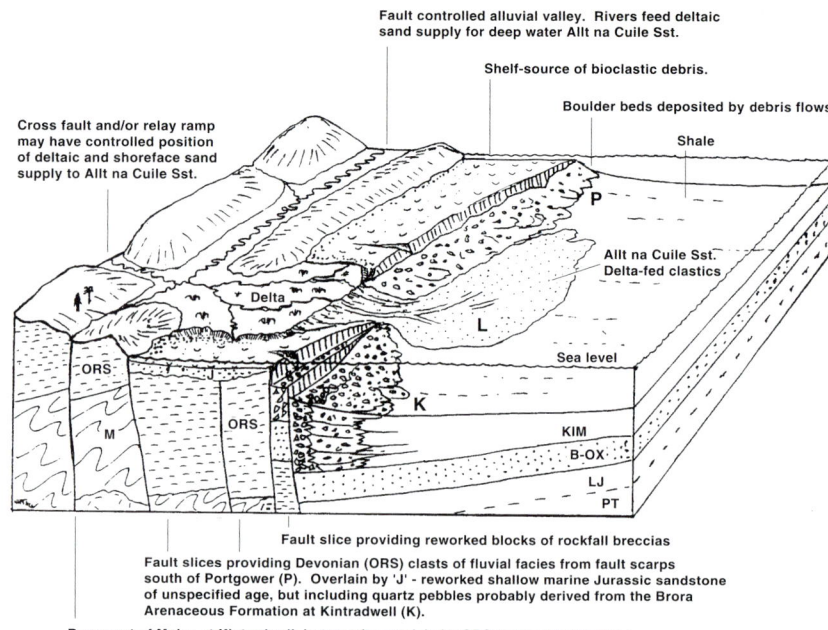

3.4 Reconstruction of the Helmsdale Fault zone in the early Kimmeridgian to show factors associated with the derivation of the Kintradwell Boulder Beds and the Allt na Cuile Sandstone.

led both by the positions of valleys and rivers draining the Scottish landmass, as well as delta distributaries and the morphology of the fault scarp. The Helmsdale Fault is not a single line of fracture, and sediment transport to deep water was probably controlled by ramps and terraces between fault segments (Fig. 3.4).

Visitors to the area should find time to read Bailey and Weir's article of 1932, which is one of the classics of early sedimentological work. It gives a fascinating overview of the research history, and correctly, through a series of logical assertions, reconstructs the palaeogeography at the time of deposition of the boulder beds. More recent studies on the area by Linsley (1972), Neves and Selley (1975), Pickering (1984), MacDonald (1985), Wignall and Pickering (1993) and Macdonald and Trewin (1993) have added detailed sedimentology and palaeontology to the picture.

Biostratigraphic work based intially on ammonites (Lee, 1925; Linsley, 1972) and more recently on palynology (Lam and Porter, 1977; Riley, 1980; Barron, 1986) has shown that a virtually complete zonal sequence is present from *cymodoce* Zone of the Kimmeridgian to *albani* Zone of the Middle Volgian (Fig. 3.2). Further information (Wignall and Pickering, 1993) has helped tie the palynological and ammonite zonal schemes.

Similar fault-controlled sedimentation of late Jurassic age occurred adjacent to the margins of many North Sea Mesozoic grabens. The best-documented offshore examples are the reservoirs of the Brae Oilfield area at the western margin of the

South Viking Graben where Jurassic is faulted against probable Devonian rocks (Stow *et al.*, 1982; Turner *et al.*, 1987). These papers contain illustrations of typical core material. Some of the Brae sequences are similar to those seen in this excursion. The main oil-bearing sandstone and conglomerate reservoirs, however, were deposited as parts of a complex submarine fan system, and the sequences differ from those developed along the Helmsdale Fault. Brief summaries of oilfields in the Brae area can be found in Gluyas and Hitchens (2003). Some fields have reservoirs with breccias and conglomerates adjacent to faults, but others are more distal with respect to source and comprise sandstones similar to the Allt na Cuile Sandstone (e.g. Miller Field).

Variation in the type of sediment supply was probably the most important factor which led to the development of different sedimentary sequences. Fan-deltas apparently supplied abundant coarse-grained material to the western margin of the South Viking Graben, which was then available for transportation and re-sedimentation in deeper water by gravity flows. In contrast, little coarse-grained material was supplied to the western margin of the Inner Moray Firth basin. The occurrence of displaced thick-shelled bivalves, colonies of the coral *Isastraea*, attached worm tubes, brachiopods and sea urchins on the downthrown side of the fault (Itineraries 1, 3, and 4) indicates that the shelf area to the west was a shallow marine, high energy, rocky platform during the deposition of most of the boulder bed sequences.

ITINERARY 3.1
Kintradwell shore

Purpose
To examine the Kintradwell Boulder Beds of the *cymodoce* Zone together with features indicative of slumping and sliding of sediments on a submarine slope. A good example of an intrusive sandstone dyke is also seen.

Access
Cars or minibuses can be parked by the road near the entrance drive to Kintradwell House. Cross the railway carefully by means of the gates provided and descend to the shore. The localities to be visited are shown in Figure 3.5. The exposures should be visited within three hours of low water.

Introduction
The Kintradwell deposits are the oldest of the 'Boulder Bed' deposits and were deposited on the downthrown side of the Helmsdale Fault. The shales, siltstones, sandstones and boulder beds of Kintradwell lie within the *cymodoce* Zone and are, from both structural and palaeontological evidence, in part the equivalents of the sandstones and breccias of Allt Choll and Allt na Cuile (Wignall and Pickering, 1993). Thus there is rapid lateral facies variation along the line of the Helmsdale Fault.

The 'shales' are dark in colour due to the presence of mud and plant debris. In terms of grain size they are mainly siltstones, and Wignall and Pickering (1993)

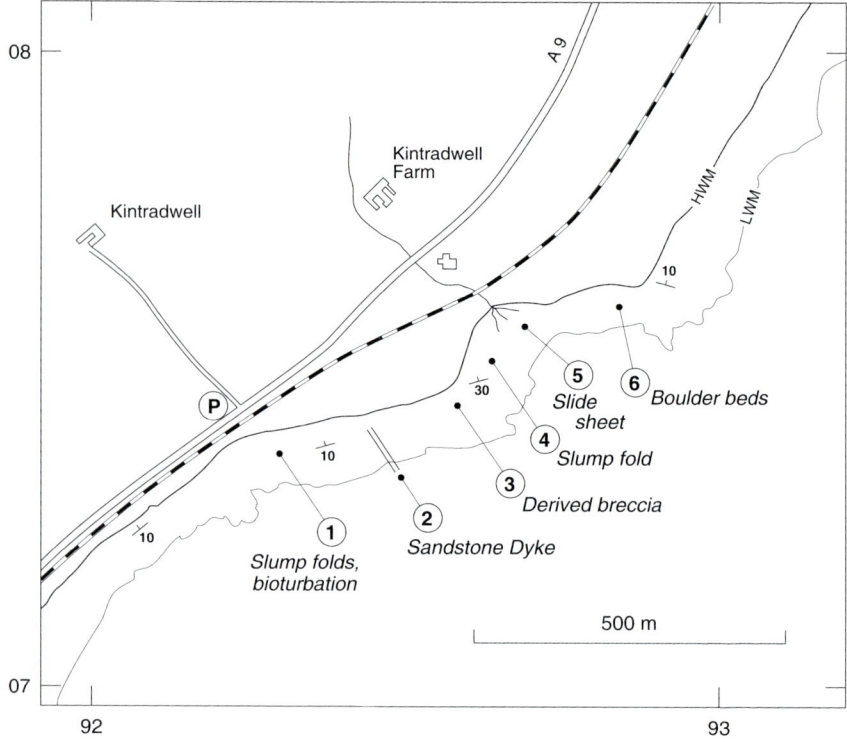

3.5 Locality map for itinerary 1, Kintradwell area.

describe them as 'paper siltstones'. They are reasonably fossiliferous, and ammonites of *cymodoce* Zone are common. The bivalve fauna includes *Liostrea*, *Buchia*, *Parainoceramus*, *Solemya*, *Nicaniella* and *Palaeonucula*, and the gastroprod *Semisolarium* is also present. Thus, bottom conditions were quite well oxygenated (Wignall and Pickering, 1993). Abundant plant debris gives evidence for the presence nearby of the Scottish landmass. Sandstone beds are sharp-based, some are graded, and they contain fragments of shallow marine bivalves as well as belemnites. Parallel lamination and ripple lamination are common and examples of convolute lamination can usually be seen. Burrows are preserved on the bases and within some beds; these may have been made by animals swept into the environment rather than by an indigenous fauna.

The soft-sediment deformation structures are varied and include both tensional and compressional features. Tensional features include pull-aparts, low angle normal faults and fractures into which sand has been injected. Compressional features include overfolds and thrusts. Large scale bedding dislocations might represent slump scars, and contorted beds and slide planes are present beneath slump sheets. Some of these structures have been described by Pickering (1983), and Roberts (1989) has published a more detailed anlysis of the soft sediment deformation features. Of particular interest is a good example of an intrusive sandstone dyke.

These sediments at Kintradwell clearly accumulated on an unstable slope at the foot of the Helmsdale Fault scarp which lay about 300 m to the west.

Locality 1. [NC 923 074]

Interbedded sandstones and shales crop out (sand cover permitting) on the lower foreshore with some prominent sandstones extending to the upper foreshore. The siltstones and shales are dark in colour, micaceous, and contain abundant plant debris. Crushed ammonites are common at some levels and benthonic molluscan faunas (see above) are present. The sandstones are generally less than 30 cm thick, display sharp bases, and are parallel and ripple laminated. Scour structures occur on bed bases and some examples of unroofed burrows may be seen. Ammonites and belemnites are also incorporated in some sand beds as clasts, and along with pebbles, acted as tools to produce prod and groove marks beneath sandstone beds emplaced by turbidity currents. Thicker sandstone beds are more massive and have deformation features such as convolute lamination, as well as faults and folds which affected the sediments prior to lithification. A strong lineation is present on some surfaces due to the soft sediment deformation; a feature well displayed on the most southerly outcrops on the shore. These beds also contain sandstone clasts, occasional rounded quartz pebbles and fairly common belemnites. The sandstone beds were formed from material swept off the shelf on the upthrow side of the fault and deposited in the deep water on the downthrow side of the fault by gravity flow mechanisms.

The quartz pebbles were possibly derived from earlier Jurassic sandstones, such as the Brora Sandstone which could have been exposed on the fault scarp at this time, and the matrix sand could be reworked from the same source. There is still some doubt as to the ages of the sandstone clasts in these beds, but most are feldspathic, and were derived from a sandstone with features of fluvial deposition. A source in the ORS is likely, the clasts having been bleached of typical red and green colours in the reducing environment of early burial (see Loc. 6 below).

Locality 2. Sandstone dyke [NC 925 074]

The sandstone dyke (Fig. 3.6) is composed of calcite-cemented sandstone and is up to 70 cm wide and near vertical. It can be traced for about 100 m over the tidal platform. This feature has been commented on by several authors since first being recorded by Murchison in 1827. It is clearly intrusive and is the result of mobilisation of sand, probably caused by earthquake shock associated with movement on the Helmsdale Fault. Pickering (1983) considered that it was intruded from below, citing evidence of upturned edges of adjacent shale beds, but this could be a compaction feature.

The reefs in the area around the dyke are worth examining in some detail. By following the strike of individual beds many discordances become apparent, which can be followed for tens of metres. Some of these may be relics of slump scars infilled with sediment; others are clearly slide planes which are frequently parallel to bedding, but change horizon to produce the discordances seen. Examples of slump folds are also present which lie on slide planes; Crowell (1961) studied these and other structures to show that derivation was from the NW. Roberts (1989), in a detailed analysis of the deformation features in this area, recognised three categories of deformation related to soft sediment movement on a palaeoslope directed away

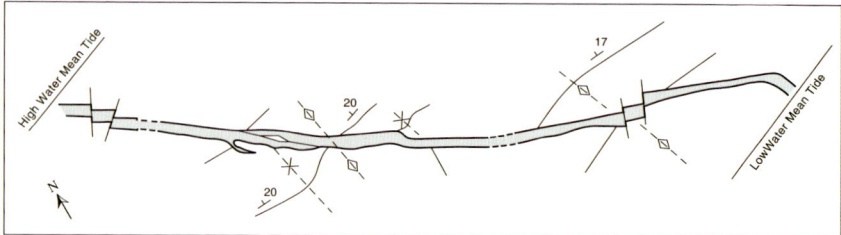

3.6 Photo and sketch plan of the intrusive sandstone dyke, Kintradwell (plan modified from Jonk, 2003).

from the Helmsdale Fault. The categories are: (1) small syn-depositional normal faults throwing down to the S or SE; (2) slump folds directed to the S and produced by movement within the wet sediment pile; (3) 'thrust' faults and related low-angle slides and extensional faults related to the movement of a gravity slide away from the Helmsdale Fault.

The most resistant sandstone reef in this area contains a great concentration of well-rounded quartz and quartzite pebbles mainly of 5–15 mm diameter, together with occasional sandstone clasts, belemnites, and shallow-water marine bivalves (Fig. 3.7). The well-rounded, spherical pebbles are also common in sandstone and boulder beds in the Allt Choll/Allt na Cuile area. They are typical of, and restricted to, *cymodoce* Zone sequences. The pebbles were rounded prior to the brief transport episode that brought them into deep water. They closely resemble those found near the top of the Brora Sandstone Member and could have been derived from that source by reworking. Likewise the matrix sand could be largely reworked from older Jurassic strata. The sporadic distribution of the pebbles suggests that the pebble source was not tapped by all events that introduced sediment to deep water. The pebbles might have been beach material channelled through gullies to the edge of the fault scarp.

3.7 Rounded quartzose pebbles and a belemnite in boulder bed, Kintradwell. Coin 28mm.

Locality 3. [NC 926 074]
Near the point a large (c.2 m) boulder is seen resting on, and deforming, a bed of laminated sandstone which varies in thickness when followed along the strike. The boulder consists of a breccia of sandstone blocks in a matrix of coarse sand with small quartz pebbles. This lithology closely resembles the breccias seen adjacent to the Helmsdale fault to the north at Allt Choll and provides evidence for early reworking of material in the Helmsdale fault zone (see Fig. 3.4). One requirement is early lithification of the breccia so that it could withstand transport. Preferential cementation is seen along fractures in the Allt Choll area and this could have been due to hot fluids rising in the active fault zone and producing near-surface lithification.

Locality 4. [NC 926 075]
The shore section is approximately coincident with the strike of the beds to locality 3 where an isolated outcrop on the foreshore, usually protruding from sand, shows an excellent slump overfold in sandstones which have a concentration of rounded quartz pebbles, together with large angular clasts of sandstone (Fig. 3.8). Clearly there were at least two sources of material, the rounded pebbles being reworked from a poorly lithified sandstone, and the angular sandstone clasts from a better lithified sandstone which probably formed the footwall of the exposed fault scarp.

Locality 5. [NC 927 076]
Cross the sandy bay and stream to the next outcrop, which is a sequence of interbedded sandstone and shale capped by a sandstone bed showing excellent convolute lamination. If exposure and seaweed permit, examine the shales at the base of the outcrop. There are two or more dislocation or slide planes present, and the shales

3.8 A Slump fold in pebbly boulder bed, Kintradwell. **B** Slide plane (at hammer head) beneath zone of deformed shale showing isoclinal folds, and overlain by a relatively undistorted sheet of sandstones and shale.

3.9 Sandstone clast showing in situ disintegration within boulder bed, Kintradwell.

between these planes are intensely deformed with thin sandstone beds distorted into flat-lying isoclinal folds (Fig. 3.9). This feature represents a slide plane beneath a sheet of sediment that slid downslope away from the Helmsdale Fault. Sand cover frequently obscures the base of this slide.

Locality 6. [NC 928 076]
At this locality impressive boulder beds are interbedded with siltstones, shales and thin sandstones. Many compaction features occur around the large boulders, and soft-sediment deformation features are common in the interbedded strata.

The boulders in the beds are angular to rounded and are composed of parallel laminated and cross-bedded white or grey sandstone. Small eroded slits in the sandstone resemble bivalve moulds, but on close examination most represent weathered-out clasts of clay flakes. Read *et al.* (1925, Frontpiece photo) considered these clasts to be of Devonian rocks, and the feldspar content is consistent with an ORS origin. However, MacDonald (1985) favoured a Jurassic age, as some undoubted bivalve casts have been reported in similar clasts in the Allt na Cuile/Lothbeg area. One large 6 x 7 metre boulder is prominent on the shore and comprises bedded sandstone with some beds showing cross-bedding and with concentrations of small green mudstone flakes; this boulder is typical of Devonian fluvial facies and unlike any of the rocks of the Triassic–Jurassic sequence. It appears that many of the clasts could have been quite poorly lithified at the time of incorporation into the boulder beds, as they show signs of disintegration and veining by the matrix muds; such features could give support to a Jurassic rather than a Devonian age for some of the clasts. However, the fluids that bleached the sandstone clasts may also have opened up joints, causing the clasts to break up during early burial and compaction. This might explain the generally angular nature of the fragments (Fig. 3.9).

The majority of clasts are derived from a fluvial sandstone of probable Devonian age. The clasts are of feldspathic sandstones and thus differ both in lithology and sedimentology from known Jurassic sandstones in the area.

ITINERARY 3.2
Lothbeg Point area

Purpose
To examine breccias, sandstones, siltstones and shales which were deposited during the initial stage of basin deepening in the *cymodoce* and *mutabilis* Zones of the Kimmeridgian. The Allt na Cuile sandstone represents a point source of sand entering the basin from a small delta fed by a river draining the Scottish landmass.

Access
Car or minibus access is from the A9 through Crackaig Farm and under the rail bridge to the Crackaig Links camp and caravan site (Fig. 3.10). Localities 1–3 can be accessed along the beach, locality 1 being about 2.5 km from the parking access. The walk involves crossing the Loth Burn. Unless the burn is in spate this may involve a shallow paddle, or some rock-hopping at locality 4. Localities 1–3 can also be reached by following a track from the A9 to the south of Loth Burn (see Fig. 3.10). Localities 3 and 6 should be visited within 2 hours of low tide. Locality 7 occurs beside the overgrown Lothbeg Bridge on the old road over the Loth Burn. The road crosses the burn about 200 m upstream from the point where the present A9 crosses the burn. Take the turning north to Loth on the east side of the burn where a car can be parked. Descend to the east bank of the burn on the south side of the old bridge. The excursion can be shortened by omitting localities 1–3 and 7, which are not suitable for large parties.

Introduction
Localities 1 and 2 display breccia-rich sequences close to the Helmsdale Fault. The precise age of the sandstones and breccias is uncertain. Brookfield (1976) obtained ammonite casts of *cymodoce* Zone age from loose blocks of sandstone, and Wignall and Pickering (1993) have proposed an early *mutabilis* Zone age for the top of the sandstone at Allt na Cuile, but most of the Allt na Cuile Sandstone is of *cymodoce* Zone age. This age is in accordance with the overall geological setting, and the sequences are thus lateral equivalents of the Kintradwell Boulder Beds (see Wignall and Pickering, 1993, Fig. 15).

The sandstones are variably bioturbated, cross-bedded, normally graded, and apparently structureless. MacDonald (1985) proposed that a possible depositional environment for these complex sequences could be a large submarine valley or canyon where tidal currents reworked the sand into dune structures, and high-density turbidity currents deposited the graded beds. The sequences clearly represent a point of considerable sand supply to the fault-margin. Possibly a river entered the sea near this point or a cross-fault channelled sand into deeper water.

Localities 4 and 5 display the best available exposures of the Allt na Cuile Sandstone of *cymodoce* Zone. They show sandstone facies at a greater distance from the

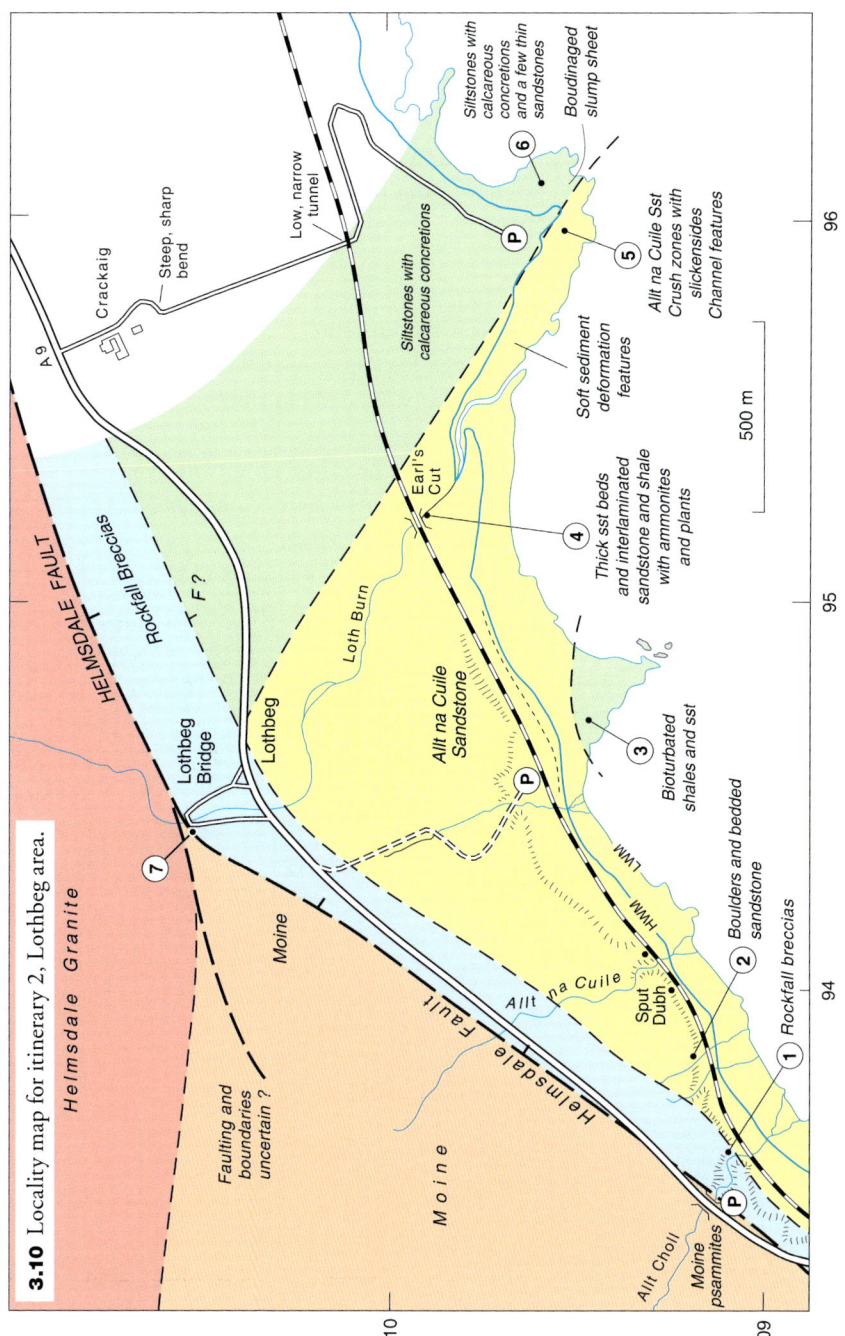

3.10 Locality map for itinerary 2, Lothbeg area.

Helmsdale fault than those of Allt Choll and Allt na Cuile. The exact relationship between the sandstones exposed at Lothbeg and in the Allt na Cuile area is not clear because of the discontinuity of outcrop. Both sequences directly overlie bioturbated mudstones (Loc. 3) and are apparently lateral equivalents on opposing limbs of a gentle northwards plunging anticline.

The sandstones at Lothbeg are better sorted and lack sandstone clasts, but occasional quartz pebbles are present. They have been made more resistant to erosion by the presence of a network of quartz cemented crush zones along the line of a cross-fault. The *mutabilis* Zone shales (Lothbeg Siltstone of Wignall and Pickering (1993)) of this area show the best examples in the Helmsdale area of black, deep-water, organic rich siltstone and shale deposition with very few thin turbidite beds included in the shales.

Locality 1. Allt Choll gorge [NC 936 091]
The NE face of the gorge is a spectacular exposure of a compound boulder bed sequence Fig. 3.11. The Helmsdale Fault lies approximately 100 m to the NW but is poorly exposed and access is difficult, particularly in summer when vegetation is high.

Poorly defined bedding dips to the SE. The clasts, up to 2 m long, are mostly friable, buff-coloured quartzose sandstones. This is a very common clast type in the *cymodoce* and *mutabilis* Zone breccias of East Sutherland. Also typical of breccias of this age at this locality are the well-rounded quartz granules and pebbles which are a common matrix component of the boulder beds, and are also seen in equivalent beds at Kintradwell (Itin. 1). It is possible that some of the clasts are reworked older Jurassic sandstones similar to the Brora Arenaceous Formation, and that matrix material of sand and pebbles is also reworked from the same source. However, it is more likely that the sandstones are extensively altered Old Red Sandstone. At the top of the cliff cross-bedded sandstones overlie the breccias. A prominent ridge at

3.11 Exposure of rock-fall breccias in Allt Choll.

the entrance to the gully consists of more strongly cemented breccia. The ridge is parallel to the Helmsdale Fault and the cementation was probably the result of fluids ascending fractures associated with the fault.

Locality 2. Stream gully to NE of Allt Choll [NC 938 092]
The SW face of the gully is composed of an interbedded sequence of breccias, sandstones and large isolated boulders of sandstone up to several metres in diameter. Other breccias and sandstones are well exposed higher up in the gully. Access is difficult in summer when vegetation is high.

The sandstones are fine-grained, thin-bedded and in places are cross-laminated and cross-bedded. The large isolated blocks of sandstone have deformed the underlying sandstones. The blocks clearly created a variable depositional topography, as the overlying sandstones have been ponded into depressions between the boulders. The overlying sandstones were deposited from relatively low-energy flows, as there is no scouring around the large sandstone blocks. This outcrop has important implications for the interpretation of depositional environments. The preservation of large angular blocks of sandstone and the lack of scouring suggests that the environment was relatively low-energy and below wave base.

Locality 3. Shoreline west of the Loth River [NC 946 095]
At this locality bioturbated sandstones and mudstones underlie the *cymodoce* Zone sandstone-rich sequence exposed at Lothbeg Point (Loc. 3), and must be cymodoce Zone or older. The bioturbated shales contain a lot of carbonaceous debris and some beds of interbedded porous sandstone. These rocks were deposited in a well-oxygenated, slowly subsiding, shallow marine basin immediately prior to the major episode of faulting and basin deepening. Subsequently, large-scale movements along the Helmsdale Fault led to a deepening of the basin, and the remainder of the Jurassic is characterised by relatively deep-marine sedimentation in a rapidly subsiding basin.

The overlying sandstone-rich sequences (Loc. 3) were deposited during a transitional deepening phase and the finely laminated black mudstones exposed at locality 4 were deposited in a deep-water, oxygen-depleted environment.

Locality 4. Loth River at Earls Cut [NC 952 099]
Large cliff exposures occur on the east bank of the Loth Burn on both sides of the railway bridge. Upstream of the bridge a sandstone bed is present in a siltstone- and shale-dominated sequence (Loth Burn Siltstone of Wignall and Pickering (1993)). At first sight, the sandstone at river level appears to be a channel, but the lower surface is concordant and the top is truncated by low-angle faulting. In the continuation of the exposures downstream of the bridge (Fig. 3.12) two 2–3 m thick sandstone beds are seen, interstratified with siltstones, shales and thin sandstones. Two large, and several small curved fault planes downthrow to the SE. They affect discrete units of sediments and probably formed soon after deposition, during failure of the sedimentary pile on a SE-dipping slope. An intensely folded slump horizon is present about half way up the cliff and numerous small syn-sedimentary

3.12 Allt na Cuille sandstones and interbedded shales in cliff at the Earl's Cut, Lothbeg.

3.13 A Ammonite with tiny encrusting bivalves and an isolated fragment of *Gleichenites*. **B** Frond of cycad from carbonate concretion found loose on beach but probably derived from the mutabilis Zone shales.

normal faults cut the sequence, many of which can be traced into bedding-parallel slip planes.

Ammonites and well-preserved plants are found associated in the dark siltstones. The ammonites indicate an open marine environment and the plants were probably derived from a vegetated delta on the upthrow side of the fault. Nineteen species of plants representing 14 genera have been described from the siltstones and shales of Kintradwell, Lothbeg Point and Crackaig Links by Van der Burg and Van Konijnenburg – Van Cittert (1984) who consider the dominant species to be typical of brackish swamps (*Gleichenites cycadina*) and freshwater swamps (*Taxodiophyllum scoticum*) of a delta. Less abundant representatives of heath, moist lush vegetation and upland forest were also recognised (Fig. 3.13).

Locality 5. Lothbeg Point [NC 960 096]
On the lower shore between localities 4 and 5 there are exposures of sandstone with some interbedded carbonaceous siltstones. These are best seen when cleaned after storms, but become algae-covered in summer. The sandstones show interesting soft-sediment deformation features (Fig. 3.14) including sand injection into pull-aparts in the dark siltstones.

The sequence underlying the white sandstones exposed near the high water mark at Lothbeg Point is generally covered in seaweed and is only exposed at low water. It is composed of interstratified sandstone and mudstone with a general increase in the sandstone proportion upwards towards high water mark. The sandstones are 2–5 m thick and are apparently lensoid. The mudstones near low water mark are bioturbated and similar to those better exposed at locality 3. The bioturbation generally decreases upwards and is replaced by finely laminated mudstone.

3.14 Features in Allt na Cuille Sandstone caused by fluidisation and injection of sand. Shore between localities 4 and 5, Lothbeg.

The sandstones at Lothbeg Point are generally friable, highly quartzose, have porosities up to 28%, and permeabilities up to 3D; thus they have potential as oil reservoirs. Some of the sandstones are clearly graded and have carbonaceous material concentrated at bed tops. The thicker sandstones at first sight appear massive, but scour features with coarser grains at the base are present. Also, cross-bedding can occasionally be observed in sandstones exposed on wave-scoured platforms.

The massive nature of most of the beds, their scoured bases and occasional grading, along with their overall stratigraphic context and interstratification with dark laminated mudstones is suggestive of deposition from high-density turbidity currents in a submarine slope or fan environment. The cross-bedding is of more problematic origin and some may be the result of localised wave or tidal reworking, but most is probably due to traction processes within submarine channels. Wignall and Pickering (1993) favour a sandy slope environment overspilling the Helmsdale Fault; they avoid use of the term 'fan' since radial organisation cannot be demonstrated.

The sandstone exposures at Lothbeg Point are crossed by numerous pale upstanding 'veins' which, on microscopic examination, are seen to be quartz-cemented granulation seams formed by microfaults; slickensides can be seen on fault surfaces (Fig. 3.15). The granulation seams are due to faulting that has hardened the sandstone over a wide zone and made them more resistant to erosion, resulting in the formation of Lothbeg Point.

3.15 Granulation seams in the Allt na Cuille Sandstone, Lothbeg.

Granulation seams such as those at Lothbeg Point, pose problems for extraction of oil from reservoirs since they divide the rock into compartments that lack effective permeable connection. They can greatly reduce the production potential of a reservoir.

Locality 6. Exposure north of Lothbeg Point [NC 961 096]

A thick sequence of *mutabilis* and *eudoxus* Zone siltstones and shales is exposed, which comprise the Loth River Shales of Lam and Porter (1977) and the Lothbeg

Siltstone of Wignall and Pickering (1993). The siltstones are organic-rich, frequently calcareous and highly fossiliferous. They contain well-preserved ammonites, belemnites, bivalves, brachiopods and drifted plants; some of the better fossil specimens occur in carbonate concretions that also yield rare fish and reptile remains. The fine lamination contrasts with the abundant bioturbation seen at locality 3, and the water depth within the basin is interpreted as having increased. Contemporaneously, the sediment–water interface was frequently, but not permanently, oxygen-depleted. The fauna at the base of the siltstone includes *Buchia*, *Parainoceramus* and *Leiostrea* and indicates moderate oxygenation. Higher in the shales oxygen levels become more depleted and pseudoplanktonic oysters attached to ammonites are the only bivalves present (Wignall and Pickering, 1993).

This organic-rich lithology is similar in appearance to the Upper Jurassic Kimmeridge Clay Formation shales which were deposited over much of the NW European continental shelf and include the organic-rich source rocks for oil in most of the North Sea oilfields. The siltstones and shales at Lothbeg and throughout the Helmsdale sequences are, however, generally more rich in terrestrially derived organic material such as plant fragments than their North Sea oil-prone counterparts, and consequently are poorer quality oil source rocks. The abundance of terrestrial material is a consequence of the basin-margin setting and testifies to a highly vegetated land area nearby.

Thin sandstones are also well exposed at this locality; these have sharp bases and tops, may be parallel-sided, or can pinch and swell across the outcrop. Some of the beds are graded and contain parallel lamination, cross lamination and waning-flow sequences of structures. The cross lamination demonstrates that these turbidite sandstones were emplaced from NW to SE by currents flowing away from the Helmsdale Fault scarp. A folded and boudinaged slump sheet of interbedded sandstone and shale is sometimes exposed at the southerly end of the exposures.

Locality 7. Helmsdale Fault zone at Lothbeg Bridge [NC 945 105]

Follow directions in the 'access' section to Lothbeg Bridge. Descend to the east bank of the burn on the south side of the old bridge. Under the bridge and upstream are exposures of shattered Helmsdale Granite within the Helmsdale Fault zone. Numerous faults are present and blocks of several different lithologies are caught up in the fault zone. Blocks of red to green laminated sandstone and mudstone, presumably from the ORS, are present and can also be seen on the west bank downstream of the bridge at river level. Extensive veins of pink and grey calcite are present in the fault zone. This exposure marks the most southerly extent of the Helmsdale Granite along the fault line. To the south of the bridge Moinian metamorphics occur on the upthrow side of the fault.

Downstream of the bridge a breccia is exposed with blocks of sandstone up to a metre in length. The boulders are generally medium-grained sandstones and contain green clay clasts and kaolinite pseudomorphs of feldspar grains. Parallel lamination and cross-bedding are present in the boulders, which are most probably of Devonian origin. The matrix is micaceous, fine sand with occasional quartz pebbles and carbonaceous material.

ITINERARY 3.3
Portgower

Purpose
To see the upward transition from sandstone-rich sequences to mudstone and boulder-bed dominated sequences and visit the Portgower giant clast locality popularly known as the 'fallen stack'. To examine changes in clast and matrix types within the boulder beds.

Access
Turn off the A9 in Portgower (car or minibus only) and park near the last houses on the lower road at the western end of the village at Craig Loisgte Place (Fig. 3.16). A track from this point runs to the cliff edge and continues down the slope to the south, reaching the shore about 200 m north of the 'fallen stack' (Loc. 4). Take care crossing the rail line. To examine the sequence in ascending stratigraphic order, walk SW along the shore to locality 2. Low tide is required; the sandstones of locality 2 should be visited within 2 hours of low water.

Introduction
The sandstones at the base of the sequence (Loc. 2) have been dated to *mutabilis* Zone by Barron (1989). They are very similar to the sandstones of Lothbeg Point but lie higher in the zonal sequence. (Itin. 2, Locs. 4–5). The sandstones pass upwards into a thin mudstone sequence which is overlain by a breccia and sandstone-rich sequence which Linsley (1972) recognised as being of *mutabilis* Zone (Fig. 3.16). These lithologies contrast with the thick *mutabilis* Zone siltstones which overlie the Allt na Cuile Sandstone at Lothbeg Point. The lack of boulder beds at Lothbeg Point

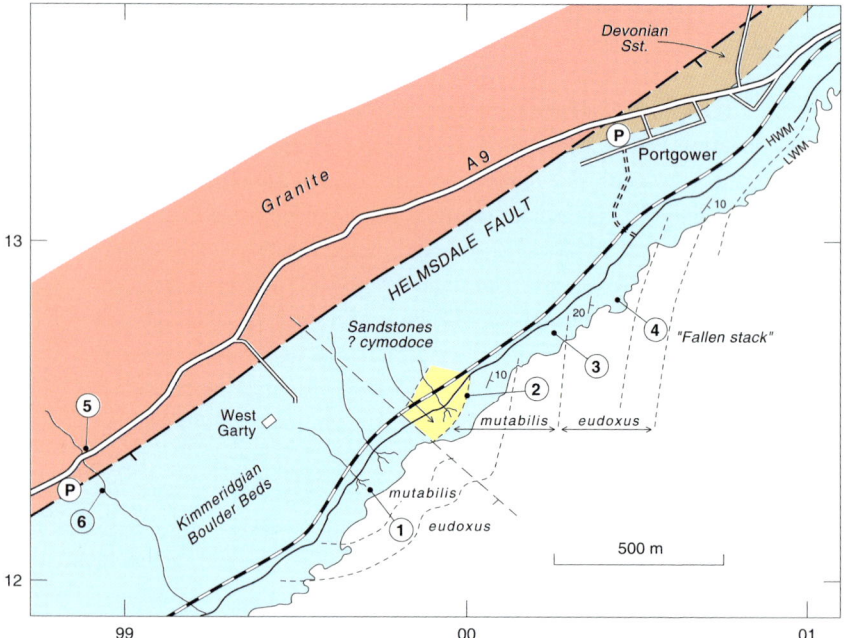

3.16 Locality map for itinerary 3, Portgower.

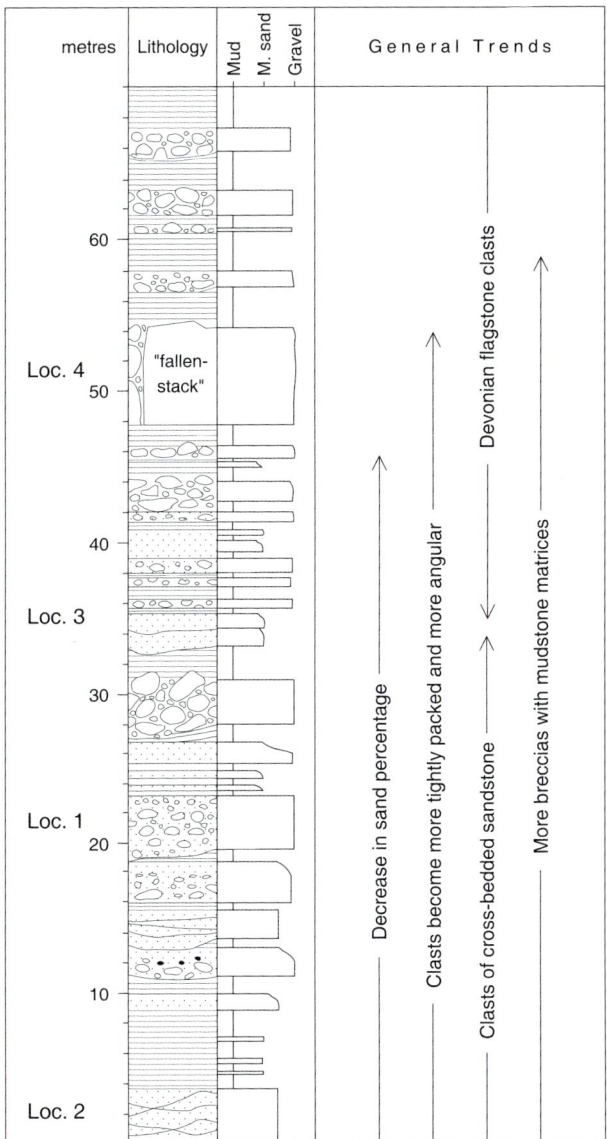

3.17 Log of boulder bed section near Portgower with general trends in lithology and clast types (adapted from MacDonald (1985)).

may also be due to greater distance from the Helmsdale Fault; thus it is implied that boulder beds did not extend far into the basin. The upward sequence from locality 2 to locality 4 reveals a change in clast types (Fig. 3.17), with clasts of trough cross-bedded sandstones of probable Upper ORS at locality 3, and Middle ORS flagstone clasts between localities 3 and 4. The misnamed 'fallen stack' of locality 4, which is in the *eudoxus* Zone, is a giant clast of Middle Old Red Sandstone flagstones.

Locality 1. [NC 997 123]
Boulder Beds in the region of locality 1 contain some sandstone clasts greater than 2 m in length. These larger clasts are red in colour in their centres but bleached to

3.18 Cross-bedding in sandstone clast within boulder bed near Loc. 1, Portgower. Lens cap 52mm.

brown and green around the margins and along fractures. The sandstone of the clasts is moderately sorted, medium-grained and trough cross-bedded (Fig. 3.18), with some evidence of distorted cross-bedding due to liquefaction. Rip-up clasts of red mudstone and occasional subrounded quartzite pebbles are present. Clasts with parallel lamination display primary current lineation. These clasts are from a sequence of fluvial origin and are probably Devonian in age rather than Permo-Triassic, since no characteristic Permo-Trias lithologies (e.g. aeolian sandstone, caliche carbonates) have been seen, and the petrographic features (mineralogy and textures) are consistent with an origin from the ORS. The large clast size and minimal matrix suggest a very local derivation, probably by rockfall. Boulder bed matrix is generally sandy with coarse shell debris, including oysters and echinoids, derived from a well oxygenated shallow shelf on the footwall side of the Helmsdale Fault.

Locality 2. [ND 000 125]
The base of the sequence is exposed in the core of a gentle anticline. The sediments are friable, medium to coarse-grained quartzose sandstones with excellent porosity and permeability. Graded sandstone beds are present with highly carbonaceous shale interbeds. These sandstones are similar to the Allt na Cuile Sandstone exposed at Lothbeg Point. The lack of any carbonate cement or carbonate bioclasts in this sandstone can be explained if the sand came direct from a delta, rather than from a shallow marine shelf. Mudstones with thin sandstone beds overlie the massive sandstone, which are in turn abruptly overlain by a breccia and sandstone-rich sequence, which marks the local base of the Helmsdale Boulder Beds. The lowest boulder beds contain clasts of pale-coloured laminated and cross-bedded sandstones, the matrix is shelly, and sea urchin spines, pectinid bivalves and oysters can be recognised. It appears that the delta had ceased to supply sand direct to deep water, and marine shelf conditions had re-established on the footwall.

From a petroleum geology perspective this exposure shows that good quality reservoir sandstones occur within the boulder bed sequence, and that there is a potential seal of shales and carbonate-cemented boulder beds to such a reservoir. The key to finding such reservoirs might be to identify possible point sources of sand (rivers and deltas) being supplied to the fault scarp in the late Jurassic.

Walk back along the shore towards locality 3 noting the clast types present.

Locality 3. [ND 003 127]
Clasts in the basal breccias at Midgarty (between Locs. 2 and 3) are predominantly light-coloured and relatively friable sandstones, and although some are similar in colour and grain size to the older Jurassic sandstones exposed at Brora, they differ in being feldspathic. These clasts are derived from a fluvial sandstone facies, probably of Devonian age. In the region of locality 3 the first clasts of typical Middle ORS flagstone facies are found. The varying clast types may reflect the progressive erosion of older lithologies on the upthrown side of the fault, but this picture is probably complicated by changes in basement geology along the line of the Helmsdale Fault where it intersects older cross-faults. Possibly the fault scarp outcrop changed from Middle ORS flagstones to Upper ORS fluvial sandstone at this point.

Locality 4. The giant clast or 'fallen stack of Portgower' [ND 004 128]
Between localities 2 and 3 a wide variety of breccias and sandstones are very well exposed and it is worthwhile to spend some time studying the textures of these beds. Exposed near the top of the beach (sometimes covered under pebbles and boulders) is a sequence of beds up to 30 cm thick and with sandstone clasts up to 10 cm, but a few reach boulder size. The interbedded shale contains sandy laminae and lenticles rich in bioclastic and carbonaceous debris. The thin boulder beds contain some excellent examples of calcitic clasts bored by the bivalve *Lithophaga*, together with oysters, and urchin spines. The latter have their long axes oriented between 130° and 150° on several surfaces. Calcite veins cut the outcrop, mostly oriented between 220° and 240°, but with some at 180°.

The famous misnamed 'fallen stack of Portgower' is a giant flagstone clast 30 m long which occurs in a breccia bed along with several other giant clasts (Fig. 3.19).

3.19 The giant clast known as the 'Fallen Stack' showing near-vertical bedding; strata in foreground show general shallow easterly dip of shales and sandy boulder beds.

Bailey and Weir (1932) measured the clast as 100 ft long. The flagstones are very similar to those exposed *in situ* in the Middle ORS Caithness Flagstone Groups throughout much of Caithness. The 'stack' shows many features typical of the Middle ORS including a dark grey laminated fish bed horizon with scattered fish scales near the landward end of the 'stack'. This lithology was deposited in a deep lacustrine environment and is described in detail in Excursion 5. Shallower water lacustrine facies include ripple lamination, some cross-bedding, and the base of one sandstone bed shows good flute moulds. Subaqueous desiccation cracks are common and some horizons of polygonal desiccation cracks are present; the latter indicating exposure and drying of the lake sediments in a playa-like environment. Thus, a full range of lithologies representing Devonian environments ranging from deep to shallow lake and exposed playa are preserved in the stack. Hugh Miller (Miller, 1854) collected Devonian (Middle Old Red Sandstone) fish fragments from this locality, and clasts of fish bed laminites still yield incomplete specimens of *Dipterus*. (However, please do not destroy such clasts in a search for the fish, which are not very well preserved; much better material can be seen at Achanarras Quarry, Excursion 5).

Early workers thought that the giant boulder was a sea stack which had fallen on its side; rather like a collapsed Old Man of Hoy. Although the massive piece of Old Red Sandstone was clearly derived from a nearby cliff, Bailey and Weir (1932) explained that it is unlikely to be a fallen stack, as it forms part of a boulder bed that is interstratified with deep-water marine shales. Bailey and Weir then correctly surmised that the giant clast was derived from the submarine Helmsdale Fault scarp. It is noticeable that in the 'stack' bed, and in many of the other nearby boulder beds, there is little matrix between the clasts. It appears that the clasts avalanched from the fault scarp into their present position, rather than being transported by any sort of debris flow process. On the basis of the 30 m stratigraphic thickness present in the giant clast, combined with the reasonable assumption that the Middle ORS in the fault scarp had a low dip, the exposed submarine fault scarp must have been at least 30 m high.

Locality 5. Westgarty [NC 989 124]
Immediately upstream of the road bridge there are exposures of Helmsdale Granite on the upthrown side of the Helmsdale Fault.

Locality 6. Westgarty [NC 989 123]
Breccias, downthrown side of Helmsdale Fault.

In the stream bed and banks both upstream and downstream of the point where the cart track crosses the stream there are poor exposures of breccias that contain large blocks of ORS lithologies. There is no evidence in this area for the continuous fault-bounded slice of Old Red Sandstone shown on the Geological Survey Golspie geological map (Sheet 103). Middle ORS fish fossils can be found in large clasts, which were thought to be *in situ* Devonian when the map was originally surveyed. In the stream between localities 5 and 6 there are poor exposures of sheared granite with a hydrocarbon-bearing vein near the contact. Parnell (1983) describes this and other minor occurrences of hydrocarbon-bearing veins in the area.

ITINERARY 3.4
Helmsdale to Dun Glas

Purpose
To examine the Helmsdale Boulder Beds, rock-fall breccias and the Helmsdale Fault zone at the northernmost end of the Jurassic exposure.

Access
This itinerary can be tackled on foot from Helmsdale Harbour (Fig. 3.20), in which case the return walking distance is 7 km if the party wishes to walk as far as Dun Glas. Low tide is required for all localities.

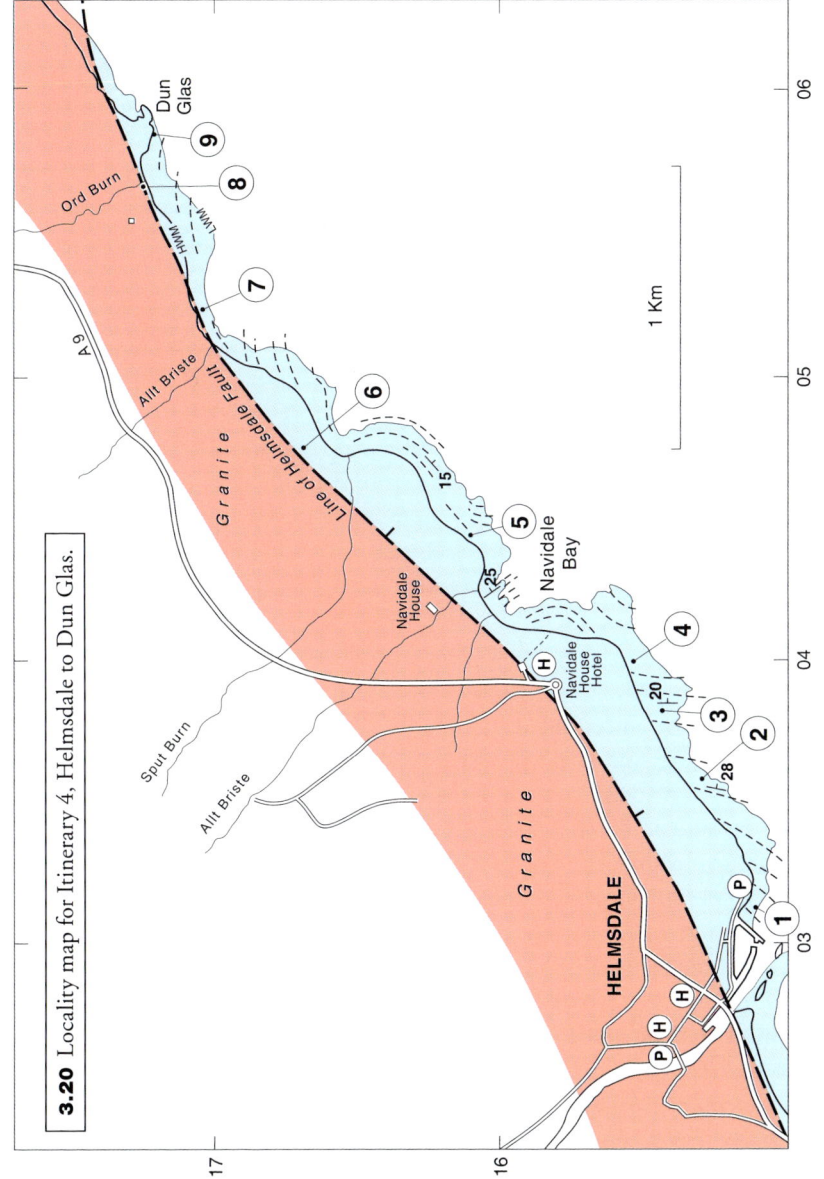

3.20 Locality map for Itinerary 4, Helmsdale to Dun Glas.

Introduction

The strata exposed on the shore range in age from *pectinatus* Zone at the harbour to *albani* Zone near Dun Glas, the latter being based on palynological comparison with the ammonite zonation (Riley, 1980; Barron, 1986, 1989)(Fig. 3.2).

The section displays a fine range of boulder bed lithologies with many interesting structures, mainly associated with deposition by matrix-poor debris flows. Clasts in the boulder beds are of Middle Old Red Sandstone flagstone facies and occasional clasts of fish-bed lithologies (see Exc. 5) yield Devonian fish fragments. *Dipterus* is the most common genus, but *Coccosteus cuspidatus* has also been found, and is indicative of a source high in the Lower Caithness Flagstone Group according to the zonal scheme proposed by Donovan *et al.* (1974).

The boulder beds are now faulted against the Helmsdale Granite, but no clasts of granite are present within the boulder beds. There has clearly been considerable post-depositional movement on the fault with downthrow to the SE. Calcite veins are oriented parallel to the fault and were formed after lithification of the boulder beds. Movement on the fault took place in Kimmeridgian times when the fault controlled sedimentation, but later movements in the Cretaceous and again in the Tertiary took place after the sediments were lithified. The boulder beds show gentle folding into anticlinal closures against the fault. These structures are well seen in the reefs of the intertidal platform, which form a protection to headlands from coastal erosion. The folds have been related to stresses set up by opposed strike-slip motion on the Helmsdale Fault (sinistral) and Great Glen Fault (dextral) during the Tertiary (Thomson and Underhill, 1993).

At the time of deposition, the active fault scarp in this area exposed Middle ORS flagstones, and the narrow shelf to the west of the fault was apparently dominated by exposed rock. A rich, shallow marine fauna of bivalves, including thick-shelled cemented forms and boring forms, colonies of the coral *Isastraea*, attached worm tubes, brachiopods and cidaroid sea urchins, lived on this shallow marine high-energy rocky shelf (Fig. 3.3).

Interbedded with debris flows derived from the fault scarp are siltstones and shales with thin beds rich in bioclasts and quartzose sand. The shales contain ammonites, belemnites and occasional fish debris. Drifted plant debris, some in the form of calcitised logs and rare reptile (crocodile, turtle, plesiosaur) remains have also been found. The deep-water environment was generally anoxic during deposition of the Helmsdale Boulder Beds; hence a bottom-dwelling fauna is absent. Even ammonites are absent from much of the sequence, a feature which Wignall and Pickering (1993) suggest was due to reducing conditions in the lower part of the water column where nektobenthonic ammonites might have lived. The general features of the fault scarp in this area are illustrated in Figure 3.21.

Locality 1. Harbour area [NC 031 151]

Entering Helmsdale from the south on the A9 cross the bridge, take the first right into Dunrobin Street and then turn right again to the harbour and park at the eastern end of the harbour. The boulder beds are exposed low on the beach and good surfaces can be seen where they have been abraded by movement of beach pebbles.

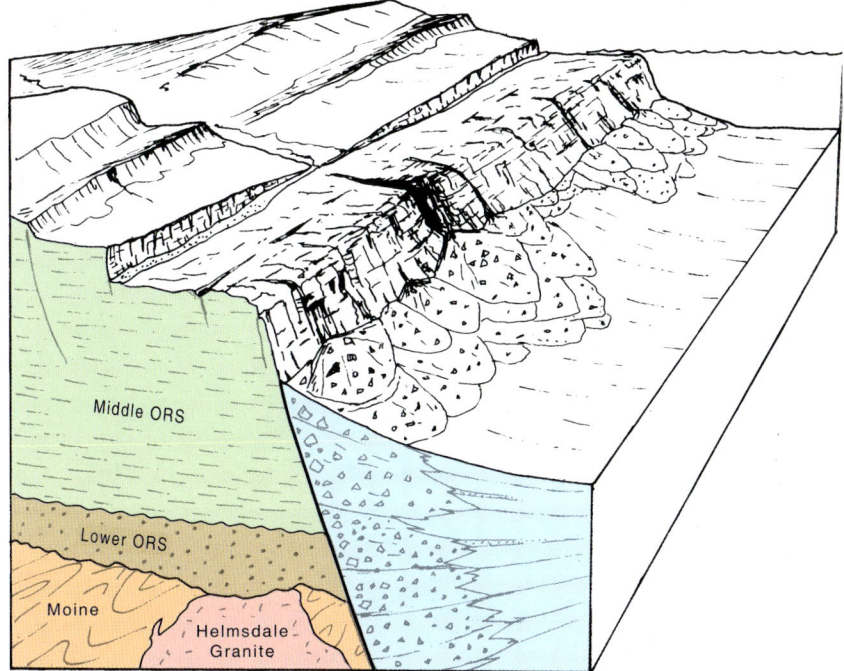

3.21 Reconstruction of the Helmsdale Fault zone in the late Kimmeridgian at the time of deposition of the Helmsdale Boulder Beds.

3.22 Typical texture of Helmsdale Boulder Beds with Middle ORS flagstone clasts in a bioclastic matrix, locality 1, Helmsdale.

Typical debris-flow textures are seen with angular to sub-angular, and also rounded, clasts of Middle ORS flagstones matrix-supported by coarse bioclastic debris (Fig. 3.22). Detailed examination reveals that rounded and angular bivalve shell debris dominates the matrix, with scattered examples of belemnites, corals, sea-

3.23 Limestone clast in the Helmsdale Boulder Beds at locality 1 bored by the bivalve *Lithophaga*, together with example showing the shape of the bivalve crypt.

urchin spines and other bioclastic debris. Clearly, there was little siliciclastic detritus available on the shelf at this point, and bioclastic debris accumulated on a rocky shelf. The variable rounding of the larger clasts reflects their derivation from shelf (rounded by wave action) or fault scarp (angular). With careful searching, clasts may be found with typical boring crypts of bivalves (cf. *Lithophaga*) (Fig. 3.23) showing that rock was exposed in the shallow marine environment.

About 150 m from the harbour some large blocks of ORS are seen, the largest measuring about 17 x 11 m in outcrop area, and others are 5–10 m in size. These blocks were probably detached from the fault scarp along joints during earthquakes, and slid the short distance (about 300 m) from the scarp to their resting place. The large block of ORS dips seawards and is inverted. Within the block examples of load structures, ripple lamination, polygonal desiccation cracks and subaqueous shrinkage cracks can be seen, all typical of the shallow lake phase deposits of the Caithness Flagstone Group (see Exc. 5).

Locality 2. [NC 035 153]
Continue along the beach, noting the variety of boulder beds and interbedded shaly lithologies. Between the shed and the concreted outfall, exposures of grey-black shales with thin, white to pale yellow sandstones up to a few centimetres thick are exposed. Sandstone beds and laminae are graded and others have ripple lamination. This thin-bedded association is similar to the so-called 'tiger-stripe' facies and has been described from the Brae Oilfield by Stow *et al.* (1982). The sand was probably deposited following storms that stirred up sediment on the shallow shelf and swept it over the fault scarp and down into deep water. There is no sign of bioturbation or a benthonic fauna and amorphous sapropelic organic material is preserved; bottom conditions thus appear to have been reducing. Drifted plant debris and ammonites may be found. Some lenticular boulder beds are seen here, demonstrating that individual flows were of small volume. Boulders in the beds exceed 1 m in diameter; indeed the clasts may be thicker than the 'bed' that contains them, and protrude from the tops of beds. Beneath the boulder beds there is frequently a zone of siltstone and shale up to 10 cm thick which has been deformed during the emplacement of the boulder bed.

Locality 3. [NC 038 154]
Proceed to about 150 m beyond the modern house where a 2 m thick boulder bed with a coarse shelly matrix can be seen. The bed wedges out down the beach, away from the Helmsdale Fault. Calcite veins, which run subparallel to the Helmsdale Fault, cut both clasts and matrix, indicating that the rock was fully cemented prior to the veining. Sandstone at the top of the bed can be seen apparently draped over boulders, but this is likely to be a compaction effect (Fig. 3.24).

3.24 Example of a boulder bed that wedges out rapidly away from the Helmsdale Fault, Helmsdale shore.

In the next dark siltstone and shale sequence, about 100 m further along the beach, some thin beds of sandstone with bioclastic material have spines of cidaroid sea urchins on their top surfaces which have their long axes aligned perpendicular to the Helmsdale Fault. The sea urchins probably lived on a rock substrate at the top of the fault scarp. A few ammonites and belemnites are found in the shales.

The interlaminated shale and sandstone in this area is rich in carbonaceous debris and shows complex syn-sedimentary faults and zones with laminae deformed into ptygmatic folds. A coarse sandstone bed is present that wedges out down the shore, and isolated boulders which slid into the muds from the fault scarp are found within shale.

Locality 4. [NC 040 155]
Near the point, there are usually good exposures of boulder beds, sandstones and shales in which lateral discontinuity of the sandstones and boulder beds can be demonstrated. About 100 m south of the point, sandstone and boulder bed lithologies occur in dyke-like bodies discordant to the bedding of the enclosing strata. Colonies of the coral *Isastraea* (Fig. 3.25) occur as boulders in some beds in this area – please do not hammer good colonies – loose material can usually be found on the shore. Logs of calcitised conifer wood are also reasonably common. Examples of beds with boulders larger than the bed thickness, boulder beds overlying disrupted shale, and small sandstone dykes are also present. From the point (Sron a' Chrochair) there is a good view northwards to Dun Glas (also known as Green Table) where the Helmsdale Fault crosses the shore and leaves the coast (Fig. 3.26).

3.25 Colony, 25 mm across, of the coral *Isastraea* from the boulder beds at Helmsdale. The colony was swept into deep water from its living position on a shallow shelf. It may have grown attached to rock on the shelf edge.

3.26 View of Dun Glas from SW of the end of Allt Briste. The Helmsdale Fault is exposed on the beach below the granite buttress at the left of the picture, and passes through the col at the back of Dun Glas.

It is a 10 minute walk back to the harbour from this point, and Dun Glas is a brisk 40 minute walk further along the shore.

Locality 5. [NC 045 161]
Continue into Navidale Bay noting the seaward plunging anticlinal closure in the boulder beds exposed on the shore beneath the Navidale House Hotel. Locality 5, at the north side of the bay beyond a ruined building, is a small cliff in boulder breccia which contains very little matrix and is interpreted to be a rockfall breccia. Rare examples of clasts bored by bivalves are present, so confirming the submarine origin of the breccia. Shales and boulder beds with bioclastic debris are interbedded with the rockfall breccias.

At the end of the cliff bedding becomes vertical against a large, subsequently fractured, block of shaly ORS some 20 m thick. The clasts in this region are all very similar ORS of local derivation, and further down the shore the typical boulder bed facies with a calcareous matrix is present. Prominent calcite veins trend at 030–050° across the outcrop, filling fractures parallel to the Helmsdale Fault. Fracturing is interpreted to have taken place after the boulder beds were fully lithified because the fractures pass straight through both clasts of ORS flagstones and Jurassic matrix, rather than going around the clasts.

Locality 6. [ND 047 167]
The reefs of boulder beds on the shore form a seaward-plunging anticlinal fold around the point and protect the point from erosion. The waterfall formed by the Sput Burn cascades over the rockfall breccia close to the line of the Helmsdale Fault. At the back of the bay a prominent rusty outcrop in a gully cutting the cliff marks

the position of a lens of fractured yellow sandstone caught up in the fault zone. This sandstone is medium-grained, and contains moulds of marine bivalves including pectinids and small oysters. Bailey and Weir (1932) considered that these could be 'Corallian' sandstones that had collapsed into a chasm along the line of the fault. Both Lee (1925) and Bailey and Weir (1932) compared this sandstone with that of the Loth area (Allt na Cuile Sst.) and in particular the Allt Choll breccia. The sandstone is certainly no older than Jurassic and is possibly of Lower Cretaceous age. Lower Cretaceous crops out on the seafloor close by, so the presence of a slice caught in the fault zone is not surprising. It is also possible that the sediment was intruded as a clastic dyke, or represents part of a fissure filling in the fault line as suggested by Bailey and Weir (1932).

Locality 7. [ND 052 171]
The Helmsdale Fault zone is exposed on the shore at this locality near the mouth of Allt Briste (Fig. 3.26). Rocks in the fault zone are intensely fractured, sheared and veined. Cherty crush veins are prominent and original lithologies are hard to recognise, but it appears that most of the fault zone consists of altered granitic material. Veins with calcite and pyrite are also present. Typical boulder beds are seen on the SE side of the fault on the foreshore. The fact that the granite is so intensely deformed when compared to the softer Jurassic provides evidence that the granite deformation took place at great depth, and that the Helmsdale Fault is an older structure reactivated in the Jurassic. Bailey and Weir (1932) considered that 'a foot or two' of white sandstone at this locality rests unconformably on Helmsdale Granite. It is, however, difficult to distinguish small fault slices, vein infills and unconformable relations in such disturbed rocks.

Locality 8. [ND 056 173]
The Ord Burn forms a waterfall at the sharp transition from the Helmsdale Granite to the Helmsdale Fault zone. The fault zone is exposed in a NE-facing cliff adjacent to the burn and mostly comprises sheared granitic lithologies. This contrasts the situation in the Garbh Allt to the south of Helmsdale where a slice of ORS is present between the granite and the Jurassic strata. In general there is little evidence for ORS lithologies in the footwall of the fault to the NE of Helmsdale. However, in the raised beach cliffs to the SE of locality 8, some probable ORS conglomerates with rounded granitic clasts and arkosic sandstones are poorly exposed. Clearly, there are many lithologies of various ages incorporated as fault slices within the Helmsdale Fault zone.

Locality 9. Dun Glas [ND 058 172]
Spectacular cliff exposures of rockfall breccias dipping steeply to the SE are exposed here (Fig. 3.27). The fault is not exposed but runs adjacent to the breccias. The steep south-easterly dip is partially due to drag effects during later movements on the fault. Note the variety of sandstone and flagstone clasts and the lack of interstratified mudstone.

Compound amalgamated breccia sequences are typical of a narrow zone lying immediately to the SE of the Helmsdale Fault. These breccias are the deposits of

3.27 Inclined rockfall breccia of Middle ORS flagstone clasts at Dun Glas, locality 9, Helmsdale excursion.

fossilised submarine fault-scarp talus slopes. The sequences of interstratified mudstone, sandstone and boulder beds which are exposed at Portgower and Helmsdale were deposited further from the fault, where settling of suspended mud particles was the background sedimentation. Occasional catastrophic rock avalanches, debris flows and turbidity currents transported coarser-grained sediment into this deeper marine low-energy environment.

Dun Glas has a capping of glacial till. If one climbs to the col at the back of Dun Glas the line of the Helmsdale Fault can be seen at low tide in the shore reefs to the NE, where the fault passes out to sea.

Excursion 4

The Lower Old Red Sandstone and Helmsdale Granite of the Ousdale area

N. H. Trewin

Purpose

To examine the Ousdale Arkose and Ousdale Mudstones and associated sandstones of the Lower ORS. The Helmsdale Granite with minor mineralisation is seen in roadside cuttings and quarries. Excellent distant views are obtained of structures in the Helmsdale Boulder Beds at Navidale. The basal beds of the Middle ORS between Ousdale and Berriedale are covered in Excursion 5, localities 1 and 2, but could be added to this excursion if desired.

Access

Drive north from Helmsdale on the A9 as far as the parking place for locality 1 in a lay-by by a small hut [NC 065 199] at the southern end of Ousdale road cutting (Fig. 4.1). Localities 2–6 can be visited on the return journey to Helmsdale. All localities are within a 9 km drive of Helmsdale, and are close to the roadside. The excursion need only take 2 hours, but can be extended to half a day if time is spent in detailed examination of the sedimentology and a search for trace fossils.

Introduction

The Lower Old Red Sandstone of the Ousdale area rests unconformably on the Helmsdale Granite, which is intruded into Moinian metamorphic rocks. In the Ousdale area, the Badbea Breccia (Exc. 5, Loc. 1) at the base of the Middle ORS unconformably overlies Lower ORS (Fig. 4.2), but further north at Sarclet only a disconformity is apparent (Collins and Donovan, 1977).

The Lower ORS of this area was previously assigned to a 'Basement Group' which Westoll (1951) suggested was of Early Devonian age. This age was confirmed on the basis of spores (Richardson 1967), and further strengthened by the greater variety of spores recorded by Collins and Donovan (1977) indicating an early Emsian age. Collins and Donovan (1977) also record the finding of *Porolepis* scales from Ousdale, which support an Emsian or early Eifelian age.

The Lower ORS sediments were deposited in a variety of alluvial environments bordering outcrop of the Helmsdale Granite. The granite appears to have been deeply weathered to form a 'grus' of granitic debris, and a precise erosive contact between the granite and arkose is not clearly seen on this excursion.

Minor uranium mineralisation is associated with the Helmsdale Granite (Tweedie, 1979), most being associated with hydrothermal and weathering alteration products. The uranium-bearing minerals kasolite, meta-autunite and meta-torbernite

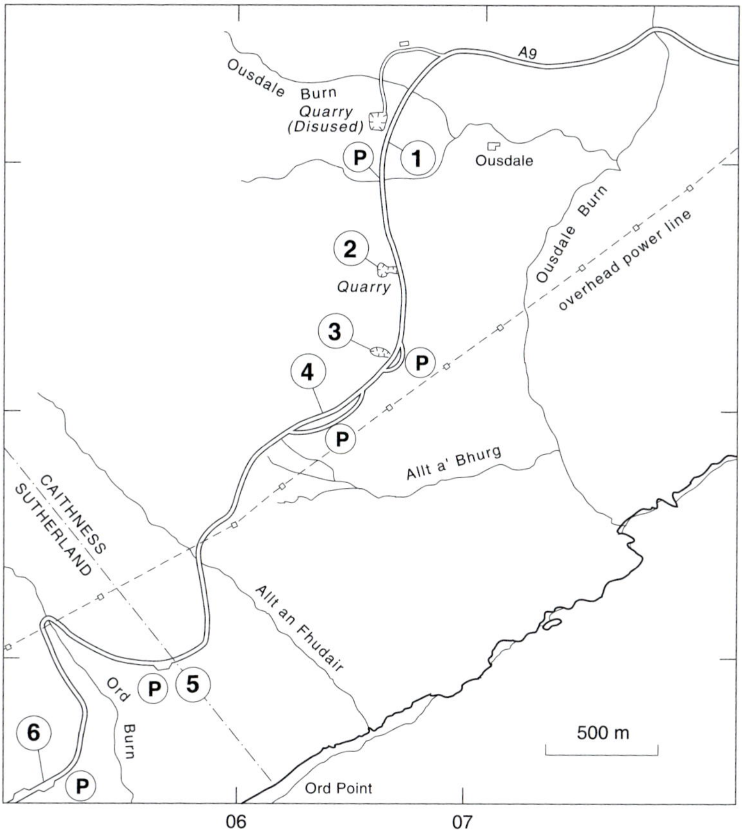

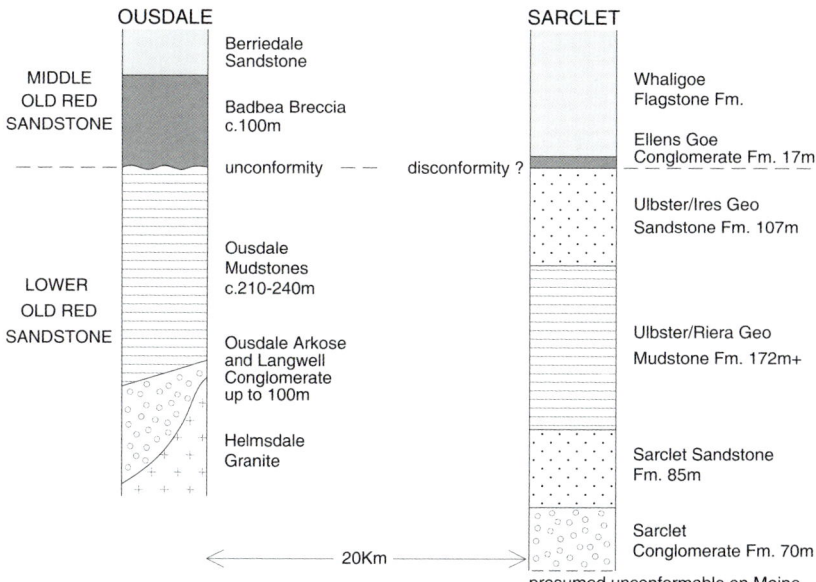

4.1 Locality map for excursion 4, Ousdale area.

4.2 Stratigraphic sections at Ousdale and Sarclet (see Exc. 5). Modified from Collins and Donovan (1977).

have been recorded from the area. Fluorite mineralisation is seen at locality 1 on this excursion.

Locality 1. Ousdale cutting [NC 066 201]
Locality 1 is a road cutting on a rather dangerous stretch of road. Park in the small lay-by (not big enough for coaches) by a small hut to the south of the cutting (Fig. 4.1) and walk northwards down the road to the cutting.

The rocks in this area are not so easy to identify as first impressions might indicate. The western wall of the cutting is mainly Helmsdale Granite, but it is considerably fractured. Thin sections reveal that at least part of this rock is a coarse arkosic sediment. The large pink feldspar phenocrysts from the granite are fresh and angular, even when incorporated in arkose, and little transport can have taken place. These rocks lie very close to the unconformity surface between the Ousdale Arkose and the Helmsdale Granite, and it is probable that deep weathering and *in situ* granular disintegration of the granite took place to form a grus.

At the northern end of the eastern side of the cutting a small fault separates massive arkose and granite from strata which are clearly bedded and represent transported arkosic sediments (Fig. 4.3). Armstrong *et al.* (1978) interpreted this feature as the unconformity surface, and even if a small fault is present, it is certainly close to the unconformity. These rocks are good examples of first-cycle sediments derived directly from granite. Pebbles of granite are present, but there is an abundance of angular pink feldspar grains derived from the granite. The beds are parallel-sided and up to about 50 cm thick; the poor sorting with granite clasts floating in a sandstone matrix indicates rapid deposition, probably by flash floods.

4.3 Ousdale road cutting on the A9 (north end, southbound side) showing the junction between sheared granite and bedded arkose. (Locality now more overgrown with vegetation.)

Some mineralisation is seen on the western side of the cutting, with purple fluorite coating fracture surfaces and forming thin veins. Patches of brown ferruginous alteration at this locality have minor uranium mineralisation.

Locality 2. Ousdale Mudstones quarry [NC 066 195]

Walk back south on the road, past the parking area and uphill until reaching the entrance to a quarry. There is a deer fence and a new metal gate at the entrance. The quarry has been disused for many years, but recently the entrance has been renovated and some excavation has taken place. The quarry is in Ousdale Mudstones (Fig. 4.2) which are dominantly dark red mudstones and shales with subordinate fine-grained sandstones and a few coarse arkose beds. The coarser beds are best seen in the north wall of the quarry; a log of this exposure is given in Figure 4.4.

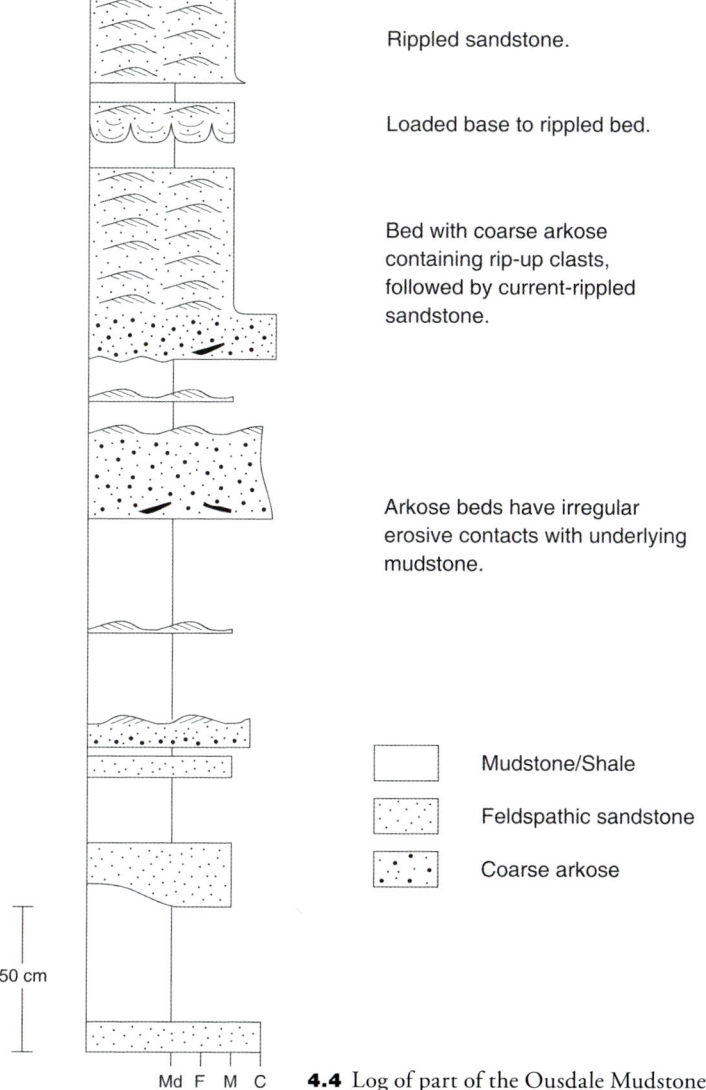

4.4 Log of part of the Ousdale Mudstones at locality 2.

4.5 Sandstone beds with erosive bases within the Ousdale Mudstones. The upper bed has a coarse arkosic base. Ousdale Mudstone quarry, locality 2.

The coarse beds have sharp erosive bases, with some scouring evident (Fig. 4.5). The coarse material is arkosic and is clearly derived from the Helmsdale Granite, and the lack of rounding of large feldspar grains is indicative of a short transport distance. The coarse arkose contains some mudstone rip-up clasts. The beds grade abruptly to a medium- to fine-grained sandstone which shows the typical small-scale trough cross-lamination produced by migrating linguoid ripples. The ripple lamination generally grades up into siltstone and red mudstone, which contains polygonal desiccation cracks.

The arkose beds appear to have been the product of individual events, probably flash floods, which swept off the exposed granite area onto the surrounding mud flats, and deposited their load of coarse granitic debris and sand. As the floods waned, rippled sand and finally mud were deposited. Subsequently the muds dried out to give polygonal desiccation cracks which can be seen moulded on the bases of some beds (best observed on loose blocks).

An interesting aspect of this locality is the relative abundance of trace fossils in comparison to most ORS sequences. They are seen on smooth mudstone lamination surfaces and bases of thin sandstone beds. Many of the trackways have low relief and are difficult to see unless the sun is shining. The most abundant trace fossil present is a small U-shaped burrow (Fig. 4.6) which should be referred to the trace-fossil genus *Diplocraterion*. These burrows were probably produced by small arthropods, and these and other arthropods were responsible for the variety of walking, swimming and burrowing traces illustrated in Figure 4.6. The arthropods inhabited the alluvial plain sediments close to the Devonian outcrop area of the Helmsdale Granite, where

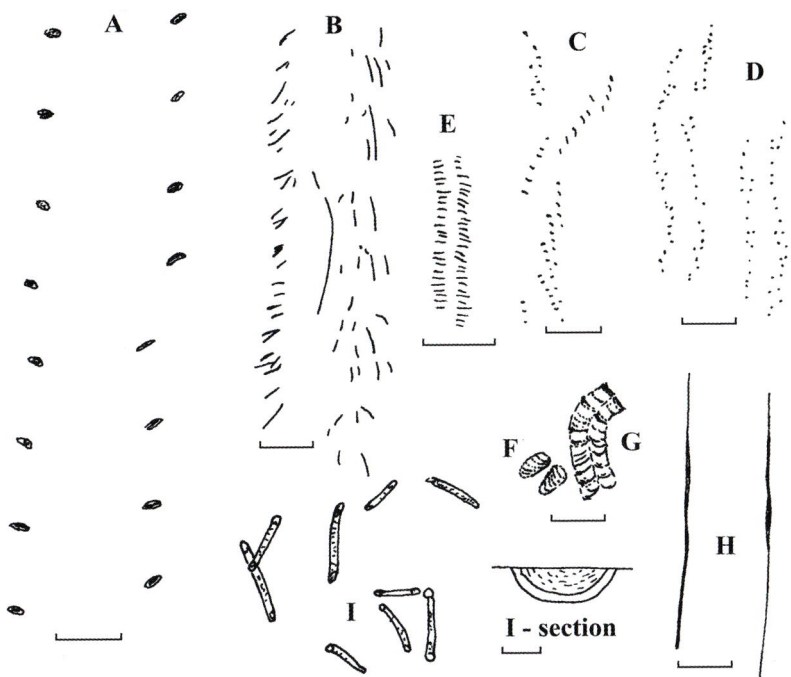

4.6 Trace fossils from the Ousdale Mudstones at locality 2. The naming of arthropod trackways A–E is tentative, and follows the work of Carroll (1990) and Walker (1985). It is probable that all these trace fossils were made by arthropods. **A** *Merostomichnites*; **B** *Allocotichnus*; **C** *Merostomichnites*, form with overlapping track series made by animal with at least six pairs of walking legs. **D** *Danstaria*; **E** *Tasmanadia*; **F** *Rusophycus*, a coffee-bean shaped resting trace. **G** *Cruziana*, a bilobed ribbon trace made by an animal ploughing through the surface. **H** *Beaconichnus*, a double groove tramway-trace. **I** *Diplocraterion*, a u-shaped burrow in plan and cross-section. Scale bars 1 cm long.

a reasonable water supply was probably available. The presence of plant debris and abundant spores at locality 3 indicates that an early Devonian terrestrial plant and arthropod community was established in the Ousdale area.

Locality 3. [NC 066 192]

A lay-by with parking area formed by a short section of the old road is situated opposite locality 3. Exposures of ORS are seen at the roadside and in a small disused and overgrown quarry where only a few exposures remain. In the quarry, a prominent bed of arkose 50 cm thick is overlain by red mudstones with sandstone beds up to 30 cm thick. Some excellent examples of current ripple lamination and small-scale convolute lamination can be seen here. The burrow *Diplocraterion* is also present at this locality. Beneath the arkose bed, three beds show good examples of dominantly planar cross-bedding in sets of 10 cm amplitude. The currents that deposited those beds, and the rippled sandstones, were flowing to the east. At the roadside a similar-looking arkose bed to the one seen in the nearby quarry is underlain by green mudstones and overlain by laminated micaceous sandstones; thus there is probably rapid lateral variation in these rocks of fluvial origin.

Locality 4. [NC 062 190]

Locality 4 is a road cutting about 1 km north of the Sutherland–Caithness boundary and 500 m north of the point where power lines cross the road. If approaching from locality 3, continue south on the main road for about 100 m and park in the next section of old road which forms a lay-by about 300 m long on the SE side of the main road; the exposures are on the NW side of the main road.

The cutting exposes sandstones and mudstones of the Lower ORS. At the SW end of the cutting, sandstones are medium-grained, micaceous and parallel laminated. Disc-shaped rip-up clasts of mudstone are present, and plant debris is abundant on some lamination surfaces. The plants are primitive terrestrial forms similar to *Psilophyton*, and *Pachytheca* (?alga) is also recorded (Collins and Donovan, 1977). Further along the cutting a sandstone unit is seen with a sharp erosive base and examples of planar and trough cross-bedding with sets to 30 cm thick; current ripple lamination is also present. Some beds contain coarse debris with pink feldspars derived from erosion of the Helmsdale Granite. In places a thin intraformational conglomerate of mudstone clasts is present at the base of the sandstone.

The sandstones were deposited in an alluvial channel and the mudstones represent alluvial plain deposits. Deposition was probably from individual floods which ripped up the mud flakes and transported the plant debris. Armstrong *et al.* (1978) considered a braided channel system to be the most likely environment. Transport direction was to the NE on the basis of cross-bedding and ripple lamination.

The combinations of lithologies, sedimentary structures and trace fossils have been interpreted in broad environmental terms in Figure 4.7.

Locality 5. Viewpoint [NC 057 180]

Stop at the lay-by and viewpoint on the Caithness/Sutherland border. On a clear day the production platforms of the Beatrice Oilfield, together with two wind turbines, can be seen some 30 km to the east. The oilfield produces from a succession containing sandstones of Early Jurassic to Callovian age which resemble the rocks seen in Excursions 1 and 2. The reservoir sandstones are overlain by Upper Jurassic shales which do not contain boulder beds. The boulder beds adjacent to the Helmsdale Fault probably only extend for a few kilometres offshore. The structure of the Beatrice Oilfield is a tilted fault block; details of the production history are given in the Geological History section of this guide. To the south Lothbeg Point and Brora stand out with the mountains of Easter Ross beyond. Tarbet Ness and lighthouse at the northern end of the Black Isle can be seen, and on a clear day the mountains to the south of the Moray Firth are visible.

Continuing south on the road, Helmsdale Granite crops out on the right before the road turns sharply over the Ord Burn, but the former lay-by at this point has been destroyed by roadworks. Granite was well exposed during road construction, but most has now been landscaped and grassed over.

Locality 6. [NC 052 175]

The view from this locality is best seen from the lay-by on the southbound carriageway, but can be see from the lay-by on the northbound side by climbing

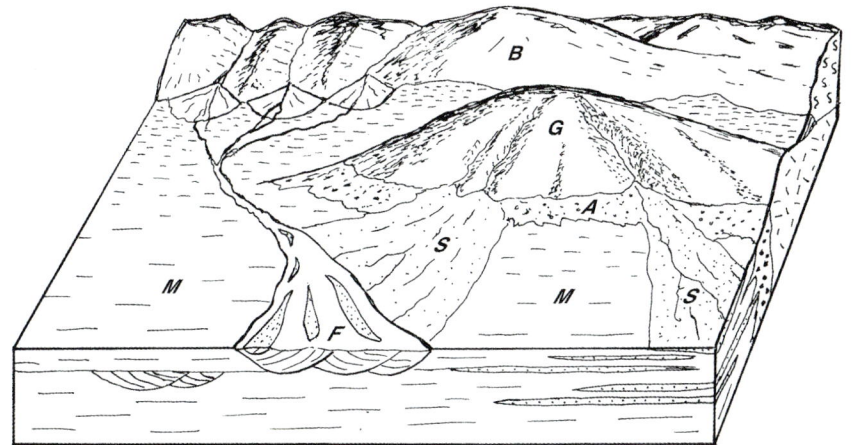

4.7 Sketch reconstruction of depositional features associated with the Ousdale Arkose and Ousdale Mudstones. Eroding granite (G) supplies material for a fringe of arkose (A), and arkosic sheetflood deposits (S) that partly cover alluvial plain mudstones (M). Fluvial channel deposits (F) are sourced from more distant metamorphic basement (B) and hence carry a variety of lithic clasts.

up the steep roadside bank. If the tide is reasonably low (and weather fine) the structures in the Helmsdale Boulder Beds on the beach at Navidale can be observed. Anticlinal folds plunging seawards abut the Helmsdale Fault. It appears that these are tectonic folds rather than depositional fans of boulder bed lithologies. Thomson and Underhill (1993) relate these folds to stress set up by opposed strike-slip motion on the Helmsdale Fault (sinistral) and Great Glen Fault (dextral) during Tertiary times. The Helmsdale Fault passes under the Navidale House Hotel (white buildings in trees near point), and trends north-eastwards to the coast at Ord Point, which is due west of this stopping place. To the SW the line of fault coincides with the break in slope between the narrow coastal strip and the hills of Helmsdale Granite and Moine metamorphic rocks.

Excursion 5

The Old Red Sandstone of Caithness

N. H. Trewin

General purpose

Localities have been selected to illustrate the variety of features of the Old Red Sandstone (ORS) of the Caithness area of the Orcadian Basin. Each itinerary would make a suitable day excursion given favourable tides, thus the localities have been arranged geographically rather than with regard to stratigraphic order. General areas of the four itineraries are shown on the Excursion Planner map.

Major features of the itineraries are listed below.

Itinerary 5.1
Locality 1 Badbea. Basal Middle ORS breccia.
Locality 2 Berriedale. Berriedale Sandstone.
Locality 3 Achanarras Quarry. Achanarras fish bed.
Locality 4 Dirlot. Middle ORS/basement unconformity.
Locality 5 Spital Quarry. Middle ORS Spital Group.

Itinerary 5.2
Locality 6–8 John o' Groats area. John o' Groats Sst., Duncansby volcanic vent.
Locality 9 South Head, Wick. Middle. ORS: Lower Flagstone Group cycles.
Locality 10 Sarclet. Lower ORS Sarclet Group.

Itinerary 5.3
Locality 11 Brims Ness. Upper Flagstone Group cycles.
Locality 12 Holborn Head Quarry. Upper Flagstone Group fish bed.
Locality 13 Pennyland Shore, Thurso. Upper Flagstone Group cycles.
Locality 14–16 Dunnet Head. Upper ORS, Dunnet Sandstone and vent.

Itinerary 5.4
Locality 17 Red Point. Unconformity ORS/basement.
Locality 18 Port Skerra. Unconformity ORS/basement.
Locality 19–23 Baligill. Basin margin deposits.
Locality 24–5 Sandside Bay. Aeolian sandstone within lacustrine flagstones.

General access

The localities are generally accessible by car or minibus combined with short walks often on boulder strewn shores, cliff tops and in quarries. Care must be taken at all times, but especially in wet weather when grass on cliff edges and rocks on the shore are very slippery.

Larger vehicles should check access and turning opportunities on minor roads to the coast and to quarries, and a little more walking will be required. Permission should be obtained to cross farmland to reach localities.

Introduction

The localities chosen from the extensive exposures of Caithness illustrate the main features of the Lower, Middle and Upper ORS in the area. The stratigraphic framework has been presented in the Geological History section of this guide. Lower ORS near Helmsdale is covered in Excursion 4 and the Lower ORS at Sarclet in this excursion (Loc. 10). Upper ORS is only seen at Dunnet (Locs. 14–16) where it is dominantly of fluvial origin but shows considerable aeolian influence. The main interest of the area lies in the Caithness Flagstone Groups and John o' Groats Sandstone Group of the Middle ORS. The general sedimentological features of this cyclic sequence are introduced below, but the visitor will find that in the field there is great variety within the cycles displayed at Brims Ness (Loc. 11), Pennyland Shore (Loc. 13), South Head, Wick (Loc. 9), Sandside Bay (Loc. 24) and Achanarras Quarry (Loc. 3). This variety reflects variations in rate and mechanism of sediment supply, subsidence, carbonate production and water depth and its chemistry. In some parts of the sequence permanent lake conditions were predominant, in others ephemeral playa lakes were usual.

The Middle ORS Flagstone Groups of Caithness and Orkney form part of a thick (c.4 km aggregate thickness) Middle Devonian sedimentary fill of this part of the Devonian Orcadian Basin (Donovan *et al*., 1974). In its broadest interpretation the 'Orcadian Basin' extended from the southern shores of the present Moray Firth through Caithness and Orkney to Shetland and Norway. In reality a great number of smaller sub-basins are present within the area with varying proportions of Lower, Middle, and Upper ORS in their fills. For further general details see the Geological History section of this guide and Trewin and Thirlwall (2002).

The flagstones are the deposits of a large ephemeral lake, the extent of which was controlled by climatic fluctuations (Hamilton and Trewin, 1988; Andrews, 2008). The resulting sedimentary deposits are cyclic in nature (Fig. 5.1), and a range of conditions from deep to shallow lake and exposed playa surface are represented. The Associations A–D of Donovan (1980) are used in this guide (Fig. 5.1) as they describe the lithologies and sedimentary structures/features seen in the field. However, this is a simplification of Donovan's scheme, and there is great deal of variation within the described Associations.

The fish beds (Association A) were deposited during the deep lake phase when water depths as great as 80 m may have been achieved (Hamilton and Trewin, 1988). The fish bed lithology is a finely laminated siltstone with clastic, carbonate and organic laminae. These deposits have been interpreted as varved sediments with an annual climatic control (Rayner, 1963; Donovan, 1980). Clastic laminae represent input from rivers in a rainy season; the carbonate laminae were deposited in the dry, warm season (summer) due to photosynthetic activity of phytoplankton in the lake, and the organic laminae represent the annual decay of the phytoplankton (autumn). Other mechanisms could have introduced clastic laminae into the lake,

CAITHNESS FLAGSTONE CYCLICITY

Environment	Lithological features		Marginal areas and regressive periods
Shallow impermanent lake, strong wind influence, frequent desiccation and high salinity in some areas.	Grey-green shales, siltstones and fine sandstones in laminae and beds of 1-100 mm; abundant symmetrical ripples; subaerial shrinkage cracks common.	**D** c. 50% (Lake centre and transgressive periods)	With increased clastic input association 'D' becomes increasingly sandy and is replaced by fluvial, lacustrine delta, shoreline and rarely aeolian sandstones.
Shallow lake with fluctuating (seasonal?) level, wave action produces rippled sediment, salinity fluctuations.	Dark grey organic-rich shale and coarse siltstone laminae in pairs averaging 10 mm; ripples and subaqueous shrinkage cracks common, rare subaerial shrinkage cracks.	**C** c. 40%	
Shallow productive lake; restrictive sediment supply; generally below wave base.	Dark grey organic-rich siltstone and shale, laminae 0.5-3.0 mm thick; only minor carbonate, rare ripples, micrograding and some subaqueous shrinkage cracks.	**B** c. 5%	Associations 'A', 'B' and 'C' become reduced in thickness or eliminated.
Deep lacustrine with some degree of thermal stratification.	Typical varved fish beds with organic, carbonate and clastic laminae.	**A** c. 5%	
		B, C, D Little clastic input	Increased clastic input

5.1 Summary of the characteristics of Lithological Associations A–D of Donovan (1980) which form the cyclic lacustrine facies of the Middle ORS of Caithness.

for example, dust storms passing over the lake (Trewin, 1991), but in general it is considered that the millimetre-scale lamination represents annual varves.

The fish found preserved in the laminites of the Achanarras fish bed (Loc. 3) include shallow-water bottom-dwelling forms such as *Pterichthyodes* which clearly drifted out into the lake as dead carcasses prior to sinking into deep anoxic water where they were preserved (Fig. 5.2). Since all fish appear to be affected, including carnivores such as *Coccosteus* and *Glyptolepis*, it is thought that mass-mortality events were responsible. This interpretation is supported by the occurrence of fish concentrations on specific lamination planes within the fish beds (Trewin, 1985, 1986).

Causes of mass mortalities might have been de-oxygenation events caused by algal blooms, or storms mixing anoxic water from the deep lake with the oxygenated surface waters or stirring up deoxygenated mud. Extreme water temperatures during hot weather can also cause mass mortalities of fish in rivers and lakes. Evidence from

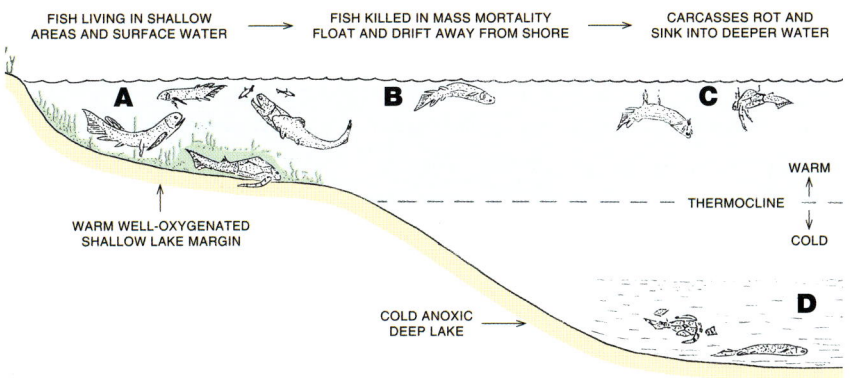

5.2 Origin of fossil fish carcasses in deep lake laminite facies. Fish lived in rivers and shallow lake areas (A) where waters were oxygenated. Periodic mortalities due to salinity crisis, or deoxygenation caused by algal blooms, lake overturn or storm mixing, resulted in carcasses (B) drifting out into the lake where they eventually decayed (C) and sank through the thermocline to be preserved in the anoxic laminites of the deep lake (D). Modified from Trewin (1986).

isotopic studies and carbon/sulphur ratios (Hamilton and Trewin, 1988) indicate that salinity in the lake was variable, and that salinity crises might have been the cause of mortalities. Disarticulation of fish carcasses is related to environmental energy and hence water depth.

The Achanarras fish bed at Achanarras Quarry has yielded fish of some 15 genera of which three can be considered abundant (*Dipterus, Palaeospondylus, Mesacanthus*) and two others relatively common (*Coccosteus, Pterichthyodes*). These five genera make up 85% of the population (Trewin, 1986). Other fish beds are seen at Holborn Head (Loc. 12), Pennyland Shore (Loc. 13), John o' Groats (Loc. 6) and Sandside Bay (Loc. 24). Further details of fauna are included in the Achanarras excursion (Loc. 3).

Immediately above the fish bed laminites beds of silt to fine sand are present at Achanarras, the sediment being transported into the deep lake by turbidity currents derived from the basin margin. However turbidites are only associated with a few of the cycles.

The fine laminites of the fish bed grade upwards into Association B which comprises organic-rich shale and siltstone with laminae generally less than 5 mm thick. Silt laminae may be graded and were probably introduced as very weak density flows. Ripples and subaqueous shrinkage cracks are only rarely found. With increased shallowing of the lake the laminae become thicker, so that Association C comprises sand/shale couplets averaging 10 mm in thickness. The shale is generally grey, and ripples and subaqueous shrinkage cracks are common and may reflect changing salinity (Donovan and Foster, 1972).

Association D represents shallow, ephemeral lake conditions as shown by the numerous polygonal desiccation crack horizons together with ripples of very shallow water origin. The shales are generally green in colour and lack organic matter due to oxidising conditions during deposition. Within this facies, sandstones of fluvial, lacustrine delta, shoreline and aeolian origin have been recognised. Near the lake margins (e.g. Red Point (Loc. 17) and Port Skerra (Loc. 18)) the sediments of Associations A, B and C become interbedded with conglomerates and sandstones of fluvial and shoreline origin, and at Dirlot Castle (Loc. 4) a rocky lake shore with stromatolites is preserved against an inlier of Moine metamorphics (Donovan, 1973).

An alternative interpretation of Associations B and C, based largely on the sections at South Head, Wick (Loc. 9) has been given by Rogers and Astin (1991) and Astin and Rogers (1991). They consider that there is a complete gradation between the lenticular (subaqueous) sand-filled cracks and the typical polygonal cracks representing subaerial exposure. They conclude that the lenticular cracks were also of subaerial origin and that they were initiated on evaporitic gypsum crystals within the sediment surface. The sand fill of the cracks is considered to have been wind-blown over the dry playa lake floor.

Whilst there are examples of sand-filled pseudomorphs, possibly after gypsum, and even rare hopper-shaped pseudomorphs after halite, the majority of cracks do not resemble gypsum crystal shapes. I prefer (Trewin, 1992) a subaqueous origin for the majority of these cracks, since they are highly compacted (indicating highly water-charged sediment rather than dried mud). The organic matter within the sediment was not oxidised by exposure, and the micaceous fine-sand fill of the cracks is not typical of aeolian transported material. Furthermore, since the horizons with lenticular cracks are repeated hundreds of times in sequence and each extends only a few centimetres in depth, it is difficult to imagine such regular and similar desiccation events. If the Astin and Rogers model is appropriate, desiccation of the lake floor was a much more frequent event than in the Donovan model which is preferred in this guide.

Further details of the fish are contained in the guide by Saxon (1975), the leaflet 'Fossil fishes of Caithness' produced by the Caithness Fossil Group, which can be found at tourist outlets, and academic papers referenced in this guide. When visiting localities in the area please remember that most of the fish-bearing localities are designated as 'Sites of Special Scientific Interest' (SSSI). Outcrops must not be hammered and bedrock must not be disturbed. At Achanarras Quarry material is found by searching the waste tips. Parts of the tips are regularly excavated by Scottish Natural Heritage to provide material for collectors to examine. At all times follow the Scottish Fossil Code, and report any unusual finds. Two new fish genera (*Cornovichthys*, *Actinolepis*) and a new arthropod (*Achanarraspis*) have been described from Achanarras in the past 10 years, showing that valuable contributions to science can be made by collectors when interesting finds are reported.

ITINERARY 5.1
Old Red Sandstone of Achanarras Quarry and the Unconformity at Dirlot Castle

Purpose
To examine the fauna and sedimentary features of the classic Middle ORS fish bed locality at Achanarras Quarry, and to demonstrate the Middle ORS unconformity

on Moine gneisses at Dirlot Castle where stromatolite coated breccias overlie the unconformity. Roadside exposures of basal Middle ORS breccias and sandstones north of Helmsdale can be briefly examined on the journey if the party is staying at Helmsdale. Alternatively localities 1 and 2 can be added to Excursion 4 (Ousdale).

Access

From Helmsdale drive north on the A9 visiting localities 1 and 2 if desired. Proceed as far as Latheron where the A9 forks left towards Thurso (Note that on older maps the A9 is the right fork to Wick, now renumbered A99). At Mybster crossroads turn left (Fig. 5.3) and after 1 km at [ND 158 528] turn right on the track beside a plantation to Achanarras Quarry (Loc. 3). Vehicles should be left at the designated parking area at the end of the plantation. Follow the 'timeline' up the track past the disused croft at [ND 153 540] to the quarry. Take care to close gates on this track. The other localities are situated near roads with easy access.

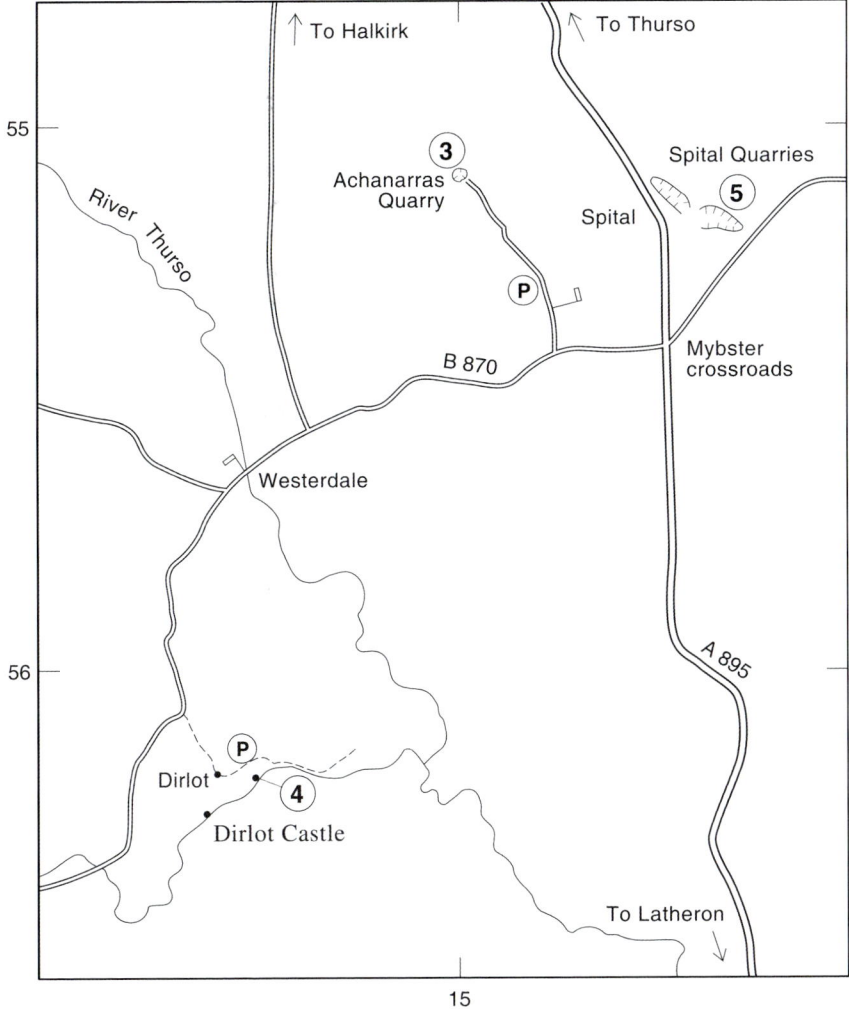

5.3 Locality map for Itinerary 5.1, Achanarras, Spital and Dirlot.

Locality 1. Badbea [ND 085 205]

The base of the Middle ORS is marked by the Badbea Breccia that rests unconformably on the Lower ORS in the Ousdale area. Examples of the breccia can be examined in small outcrops along the signposted path to Badbea historical village from the lay-by on the A9. The road cutting 200 m to the north of the lay-by also exposes this breccia. The rock contains material ranging from sand to gravel size. Most of the material consists of angular grains of intrusive acid igneous rocks and much is interpreted as having been derived from the Helmsdale Granite. Some clasts are rounded, and these tend to be pebbles of metamorphic quartzites that have a longer transport history.

Locality 2. Berriedale Cutting [ND 095 210]

The next road cutting on the journey north is 1 km beyond the Badbea cutting shortly after the power lines cross the road. Park on the right before entering the cutting. The cutting exposes Middle ORS Berriedale Sandstones. The sandstones are generally medium-grained, red and arkosic. Laminae of coarser granitic debris are also present and the Helmsdale Granite is a probable source. The sandstones occur in beds up to 1 m thick, but are generally around 0.5 m. Very little interbedded shale is present and the beds are almost exclusively parallel laminated, the lamination being due to grain-size variations. Beds appear to have been deposited rapidly, possibly as overbank deposits from large floods or as sheetflood deposits.

Locality 3. Achanarras Quarry [ND 150 544]

Follow the instructions given above in 'access' and enter this disused quarry which is situated in an exposed position on Achanarras Hill. This quarry was opened in about 1870 as a farm quarry, and was later worked by the Thurso Flagstone Company in the first decade of the 20th century. In 1959–61 it was worked for roofing slates, and minor working took place around 1970 (MacFadyen, 1992). The fish bed was first exposed in 1891 and the fauna became well known, attracting workers such as R.H. Traquair, D.M.S. Watson, E.I. White and T.S. Westoll. The fish bed succession and lamination was first studied by Rayner (1963) who proposed that the lamination represented annual varves. The quarry was drained by means of a siphon in 1980 for a study of the sedimentology and fish distribution within the bed (Trewin, 1986).

The old quarry buildings were taken down recently for safety reasons, and remains of old bogies and rails used to push waste from the quarry to the tips were removed. There is now a shelter for visitors that was opened in 2008, together with interpretation boards illustrating the mode of preservation of the fish, and providing illustrations of the fossil fish together with reconstruction drawings. Under normal circumstances the quarry is flooded and the fish bed outcrop is mainly under water, but exposures (not to be hammered or disturbed) may be visible at the north of the quarry. The exposed quarry face consists of laminated siltstones with clastic and dolomitic laminae on a millimetre scale, which are interbedded with siltstone beds up to 45 cm thick. The main features of the succession in the quarry are summarised in Figure 5.4. The siltstones (Fig. 5.5) contain rip-up clasts of the laminites, have

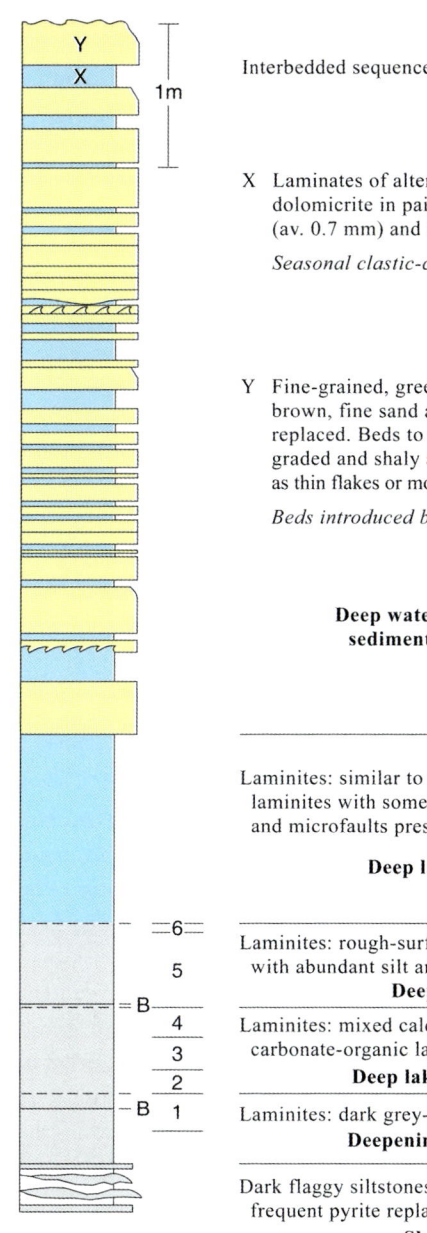

Interbedded sequence of lithologies X and Y.

X Laminates of alternating quartzose silt and dolomicrite in pairs generally 0.5-1.5 mm (av. 0.7 mm) and of even thickness.
Seasonal clastic-carbonate lacustrine varves.

Y Fine-grained, green-coloured massive beds weathering brown, fine sand and silt now extensively dolomite replaced. Beds to 45 cm, sharp based and occasionally graded and shaly at tops; beds contain rip-up clasts of X as thin flakes or more rarely as folded sheets of laminite.
Beds introduced by low density turbidity currents.

Deep water regressive phase, increased sedimentation rate due to turbidites.

Laminites: similar to above, smooth-surfaced clastic-dolomicrite laminites with some organic laminae; pull-apart structures and microfaults present.
Deep lake, continued regression.

Laminites: rough-surfaced, micronodular, dolomite-rich laminae with abundant silt and frequent organic laminae.
Deep lake, start regression.

Laminites: mixed calcareous and dolomitic, with bundles of carbonate-organic laminae, low silt content.
Deep lake, maximum transgression.

Laminites: dark grey-black, mainly clastic-organic, minor carbonate.
Deepening water, lake transgression.

Dark flaggy siltstones with paler silty laminae and isolated ripples; frequent pyrite replacement; plant debris.
Shallow lake, nearshore.

5.4 Log of section at Achanarras Quarry. Modified from Trewin (1986).

sharp erosive bases and graded tops. These beds are interpreted as turbidity current deposits of a deep lake phase and were derived from the NW (Hamilton and Trewin, 1985). Fish are not present, or possibly very rare, in this part of the succession, but a few drifted plant fragments are found.

The interbedded laminites of dolomicrite and quartzose silt are interpreted as seasonal varves. By counting these varves it is estimated that a turbidity current event

5.5 Section of base of siltstone bed resting on laminite. Siltstone bed contains rip-up clasts of laminite and the laminite consists of alternations of silt (dark) and dolomicrite (pale). The siltstone was emplaced by a turbidity current flowing downslope into the deep lake. The laminites deformed plastically beneath the turbidite; a compacted shrinkage crack produced the offsets in the lower part of the laminites in the photo. Scale bar 10 mm.

took place on average every 70 years. At some levels bedding surfaces of the laminite above the fish bed display patterns of lenticular or arcuate cracks. In cross section minor laminar displacement is seen across the cracks. These features appear to be compacted subaqueous shrinkage cracks. The origin of these and similar cracks is discussed in the introduction to this excursion and in notes on locality 9.

The fish bed laminated siltstones are very variable in texture and composition (Fig. 5.6). Individual laminae and groups of laminae are continuous throughout the quarry area, thus fish-bearing slabs can be matched to their original position in the fish bed by comparison with a collected rock section of the entire fish bed. The results of positioning over 1,000 fish are shown in Figure 5.7, which illustrates the distribution of fish within the fish bed. Details of this study have been published elsewhere (Trewin, 1986) but the major points that influence collecting in the quarry are given here. The lungfish *Dipterus* (Fig. 5.8) is common and best preserved at the base of the fish bed (Faunal Division 1) in dark, smooth, finely laminated siltstones which split easily into thin sheets. *Dipterus* is abundant on some lamination planes that represent mass mortalities of fish. A second concentration of *Dipterus* at the top of the fish bed (Faunal Division 6) occurs in coarser textured, dolomitic laminites where the fish are not so well preserved. This fish appears to have been the first to colonise the lake and the last survivor, and was probably the most tolerant of adverse conditions of oxygenation and salinity.

The central part of the fish bed (Faunal Divisions 2, 3 and 4) contains fine-grained calcareous laminites and some bundles of coarse dolomitic laminae. The

5.6 Cut and acid-etched section showing the lamination typical of the central part of the Achanarras fish bed. White laminae are dolomitic. Scale bar 10 mm.

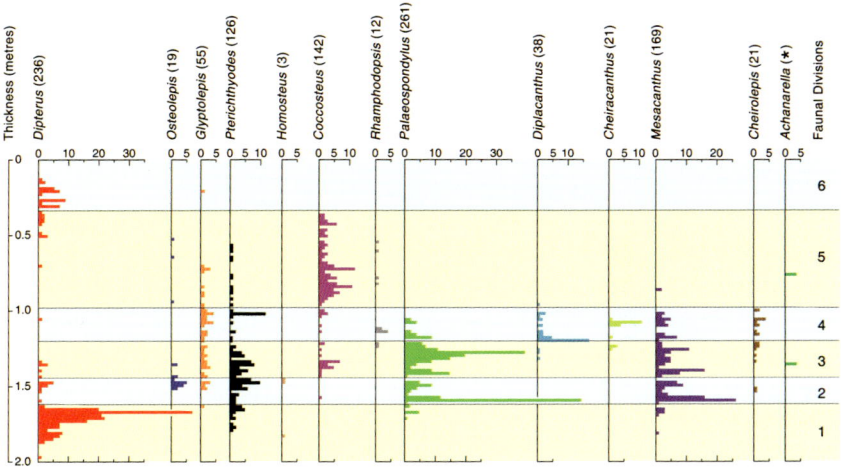

5.7 Distribution of fish in the Achanarras fish bed, together with subdivision of the bed into six faunal units. Based on the positioning of over 1000 specimens by laminite-pattern matching. See Trewin (1986) for further details.

greatest variety of fish occurs here, and *Palaeospondylus*, *Mesacanthus* and *Pterichthyodes* (Fig. 5.8) are the genera most likely to be seen. *Pterichthyodes* was a bottom-dwelling fish with a ventrally situated mouth, and eyes on the dorsal surface. This fish was first described from Cromarty by Hugh Miller (1841) in *The Old Red Sandstone*, a book that inspired many people to study geology and the fossil fishes of northern Scotland. Carcasses of *Pterichthyodes* frequently disintegrated due to decay during transport and are preserved incomplete. *Mesacanthus* and *Palaeospondylus* are frequently found together, probably because they are of similar size and carcasses

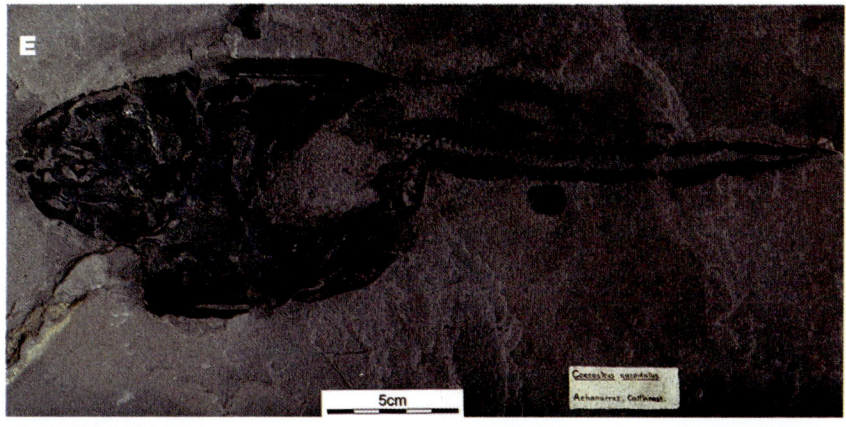
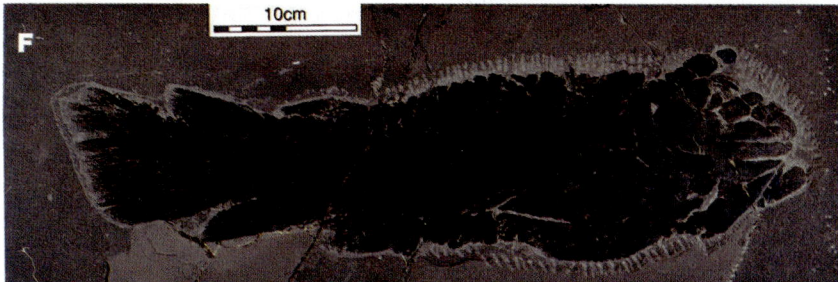

5.8 Fish from the Achanarras fish bed at Achanarras Quarry. Note difference in scales.
A *Palaeospondylus gunni*; **B** *Pterichthyodes milleri*; **C** *Dipterus valenciennesi*; **D** *Cheirolepis trailli*; **E** *Coccosteus decipiens*; **F** *Glyptolepis paucidens*; **G** *Mesacanthus*.

drifted to the same areas. The small acanthodian *Mesacanthus* may have been a shoal fish, feeding on small organisms and phytoplankton. *Palaeospondylus* is only common at Achanarras and despite much speculation there is no agreement on the relationship of this vertebrate to other Devonian forms. It has been suggested that it is a larval form, and *Dipterus* is favoured as the adult form by Thomson *et al.* (2003), but Newman and den Blaauwen (2008) provide information to the contrary. *Coccosteus* and *Osteolepis* have also been suggested as the adult form, but supporting evidence is lacking. However, the distribution of *Palaeospondylus* through the fish bed (Fig. 5.7) does not show any obvious associations with other fish, and it is likely that this fish was a chance introduction to the lake along with the other genera, and for any clues to its origin, one must look elsewhere.

The larger acanthodians, species of *Diplacanthus* and *Cheiracanthus*, and the actinopterigian *Cheirolepis* also occur in the central part of the fish bed but are scarce. They are mainly represented by full-grown specimens that probably migrated into the lake in the adult state. Scarce but large *Dipterus* also occur in this part of the fish bed. The largest predator present is *Glyptolepis*, which grew to over a metre in length and was capable of swallowing full-grown individuals of all the other species present. One *Glyptolepis* has been found, which died trying to swallow a smaller individual of its own kind, and another contained a *Coccosteus*. *Glyptolepis*, with a concentration of fins in the caudal region, was probably a lurking predator similar in habit to the modern pike.

One feature of the preservation in parts of the fish bed is that soft tissues are preserved as dark carbonaceous shadows. A few *Coccosteus* specimens show skin outlines over the dorsal fin, and the agnathans are similarly preserved. *Achanarella* (Newman, 2002) is abundant on at least two bedding surfaces, but *Cornovichthys* (Newman and Trewin, 2001) is only known from two specimens. Careful examination is required to spot these fish. The coarser laminites towards the top of the fish bed in Faunal Division 5 have a rough surface texture due to the micronodular nature of the dolomite present. The predatory arthrodire *Coccosteus* (Fig. 5.8) is most frequent at this level.

Underlying the fish bed are flaggy siltstones with isolated ripples and moderately abundant plant debris. This part of the sequence is not likely to be visible unless the lake level in the quarry is exceptionally low and the lower bench of the quarry working is visible. These rocks were deposited in shallow water at the start of the major lake transgression which resulted in deposition of the overlying fish bed laminites in deep water (Fig. 5.4).

Collecting

Fish fossils are found by searching in the tips of quarry waste and carefully splitting likely blocks of the fish bed laminates. Many people visit the quarry and good specimens are rarely found by wandering about looking at surface material. Before you start searching, take time to look at the illustrations of fossils in the display at the shelter: it is easier to find something when you know what you are looking for. On finding a specimen do not attempt to trim the slab and risk a breakage; have it sawn to shape later and prepare the fish carefully with a small hammer and chisel, mechanical engraver, or air-abrasive method. Broken material can be carefully repaired and glued back together. The material is not suitable for acid preparation. Do not take duplicate material you do not need, and leave unwanted material at the shelter for others. Any exceptional finds of new or rare genera (see Fig. 5.7) should be notified to the Royal Museum of Scotland, a local museum or University museum curator. One surprising absentee from the faunal list is *Gyroptychius*, which is common in the equivalent Sandwick fish bed in Orkney; please report any finds of this fish, or other unusual specimens.

Locality 4. Dirlot [ND 131 490]

From Achanarras quarry return to the B870 and turn right to Westerdale. Cross the River Thurso at Westerdale and continue straight along the minor road. After

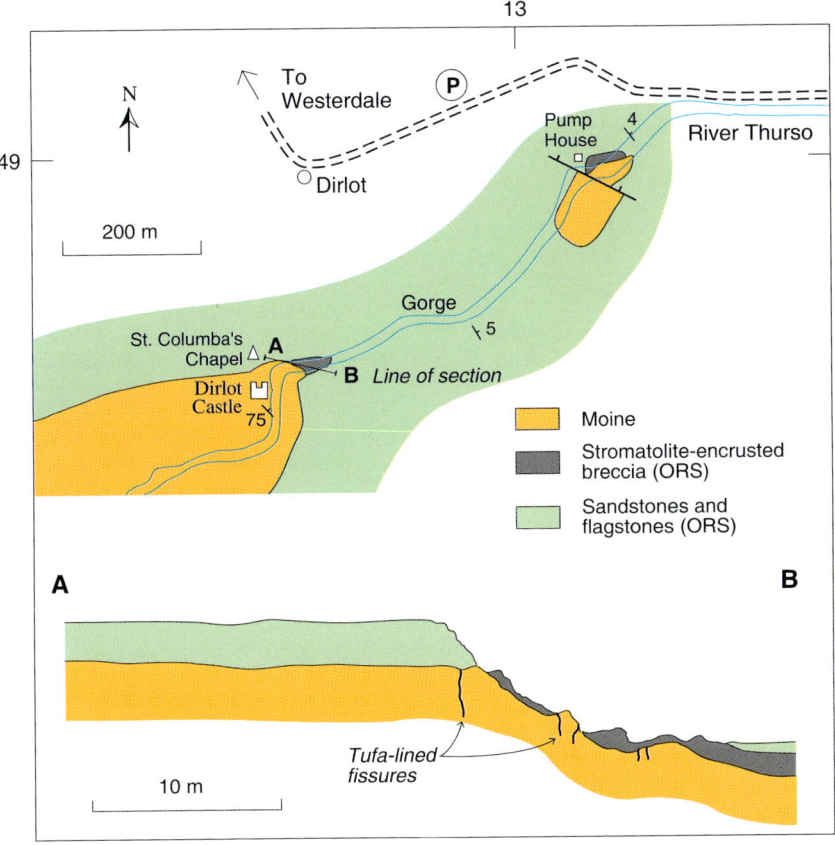

5.9 Map and sketch section at locality 4, Dirlot Castle (Modified from Donovan 1973).

a mile take the track to the left through the abandoned sand and gravel quarry. Continue a further hundred metres past Dirlot and park off the road at a small overgrown sand quarry (Fig. 5.9). Walk across the field towards a ruin on the far side of the Thurso River and find the small pumphouse at [ND 131 490]. The Thurso River is a famous salmon fishing river and parties should give priority to fishermen at this locality.

Immediately downstream of the pumphouse there are exposures of breccias which rest on a small inlier of Moine schists (Fig. 5.9). Upstream, gently dipping flagstones of the Middle ORS form the side of the gorge. The breccias which overlie the unconformity contain clasts from pebble to boulder size, some of which are coated with algal stromatolite (Donovan, 1973). The stromatolite coatings vary from thin (1–10 mm) rusty brown ferroan dolomicrite coats on smaller clasts, with larger clasts coatings up to 10 cm thick which show laterally linked hemispheroids 5–15 mm in diameter of algal or cyanobacterial origin (Fig. 5.10). Flakes of stromatolite material also occur in the matrix of the breccia. Towards the top of the exposed breccia a sandy matrix is present and the clasts are not coated by stromatolite.

Upstream from the pumphouse the river generally follows the strike of the ORS flagstones, which here consist of fine-grained sandstones in beds generally less than

5.10 Domed stromatolite grown on the surface of a boulder of Moine schist from the Dirlot breccia; matrix contains flakes of stromatolite boken from the surfaces of other clasts. Coin 25mm.

20 cm thick, interbedded with laminated shale and siltstone. Subaqueous shrinkage cracks are common and polygonal desiccation cracks are also present. The sandstones display parallel and ripple lamination and some loading features. These deposits formed in a shallow lake phase (see Introduction).

Unless the river is very low it is not possible to follow the bank of the river through the gorge, and it is easiest to walk along the gorge top and round the back of the old churchyard and St Columba's Chapel and descend to the river pool (The Devil's Pool) at Dirlot Castle [ND 125 485] (Fig. 5.11). The castle rock and outcrops beneath the churchyard wall are of Moine metamorphic rocks. Semi-pelites with gneissose banding defined by micas are prominent and quartz segregation veins are present. The foliation dip is generally steep, but a few folds with sub-horizontal axes are present. Ptygmatic folding is seen in quartz veins within the folds. Veins of aplite have intruded the gneisses and cut the foliation.

Immediately downstream of the churchyard wall over the deep pool the unconformity suface can be seen about two-thirds of the way up the face of the gorge. The surface is sub-horizontal and is overlain by about 0.6 m of conglomeratic sandstone with both angular and rounded pebbles. Sandstones with low-angle cross-bedding overlie the conglomerate.

Donovan (1973) described a network of tufa-coated fissures up to 2 m deep in the Moine schists beneath the unconformity surface. Return via the cliff top and scramble down to the riverside at the downstream end of the cliff beside the tail of the pool. The unconformity surface dips more steeply eastwards and a coarse breccia is present, apparently banked against the sloping unconformity surface. Stromatolite coatings are again present on pebbles and boulders in this breccia.

5.11 View downstream at the Devil's Pool, Dirlot Castle. The unconformity between Moine and Middle ORS is present in the cliff to the left, largely covered by vegetation.

The general environment of the tufa-coated fissures and stromatolite-coated breccias is consistent with an origin as beach or scree deposits fringing a lacustrine shore. The small inliers of Moine at Dirlot were small islands within the lake area. Stromatolite coatings formed in shallow lake waters subject to seasonal changes in water level, salinity, carbonate content and temperature; coatings are thickest on the largest, most stable clasts, which were not moved by wave action on the shoreline. Wave action and/or desiccation was probably responsible for breaking coatings from local surfaces to provide the stromatolite flakes present in the breccia matrix.

Donovan (1973) discussed the origin of the tufa-lined fissures in the Moine and considered that the laminated tufas might have been deposited by umbrophile algae. However, the sparite cements are more likely to have been inorganic in origin, formed as beachrock cement from wave splash, capillary action or the interaction of local fresh, weakly acidic groundwater with the warmer, more saline lake waters. Grey sandstones with low angle cross-bedding overlie the breccias which Donovan (1973) associated with the invasion of the lake area by fluvial conditions from the west.

Locality 5. Spital Quarry [ND 171 541]
Return to Mybster Crossroads and turn left to Spital (Fig. 5.3). After 1 km turn right off the A895 in Spital at the bend near a telephone box and drive into the quarry of A. & D. Sutherland Ltd. The quarry is in Spital Group flagstones of the Middle ORS, which lie stratigraphically above the Achanarras fish bed. The quarry is worked for roadstone, bulk aggregates and flagstones which are prepared at the quarry. Permission to visit should be obtained from the quarry office.

Lithologies are mainly laminated dolomitic siltstone. Lamination is on a millimetre to centimetre scale and is extremely regular. The surfaces of flagstones show excellent examples of subaqueous shrinkage cracks and small (1–4 mm) bumps which are due to micronodular dolomite. Fossils are very scarce at this locality, but *Dipterus*, *Dickosteus thrieplandi* and *Trewinia magnifica* (Janvier and Newman, 2005) have been recorded.

ITINERARY 5.2
Old Red Sandstone of John o' Groats, Wick and Sarclet
John o' Groats
Purpose
To examine the John o' Groats Sandstone Group at the top of the Middle ORS succession, and see the volcanic vent at Duncansby Ness.

Access
The localities (Fig. 5.12) can all be visited on foot (8 km walk) from the car and coach park at the end of A99 at John o' Groats, or the excursion can be split in two and transport (car or coach) taken to the parking area at Duncansby Head via the turning off the A99 opposite John o' Groats post office. Binoculars for geology (and the birds) are useful at Duncansby Head. Low tide is required for the John o' Groats foreshore.

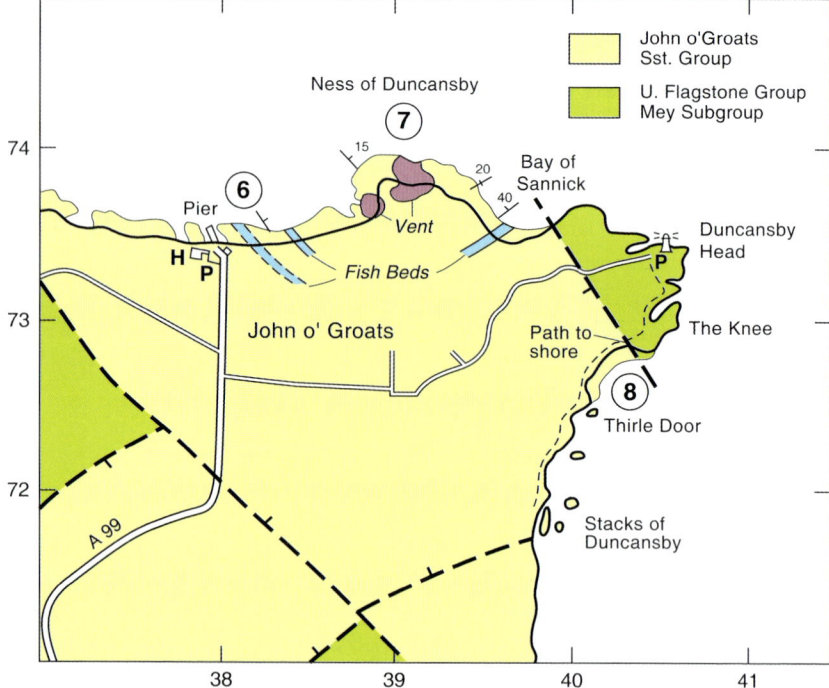

5.12 Locality map for Itinerary 5.2, John o' Groats area.

5.13 Shore to the east of John o' Groats harbour, Red fluvial sandstones with thin-bedded greenish lacustrine strata that include the John o' Groats fish bed.

Locality 6. John o' Groats [ND 380 735]

Park in the tourist car park, walk past the 'Last House' and examine the rocks immediately east of the harbour. Red sandstones dominate in beds up to 50 cm thick that display trough and planar cross-bedding, parallel lamination with primary current lineation and ripple lamination. Soft-sediment deformation features caused by water escape are frequent. These rocks are of fluvial origin and formed on a low angle, broad alluvial fan characterised by shallow channels. Further details of these rocks and comparisons with the equivalent Eday Group in Orkney are found in Astin (1985).

Periodically the alluvial fan was transgressed by lake waters, and lacustrine sediments were deposited. Some 100 m east of the harbour green to grey thin-bedded sediments represent ephemeral lake conditions and display numerous polygonal desiccation cracks, beds with wave and current ripples, and lenticular cracks of subaqueous origin (Fig. 5.13). Thin developments of dark grey laminated siltstone with pale concretionary carbonate record periods of permanent, deeper lacustrine conditions, and several of these units contain fish remains, usually as scattered fragments. One bed is the John o' Groats fish bed that has yielded *Tristichopterus alatus*, *Pentlandia macroptera*, *Microbrachius dicki* and *Watsonosteus fletti*, a fauna also typical of the Eday Group of Orkney. The fish beds are truncated by a fault and are only visible low on the shore.

Continue along the shore, noting the variety of sedimentary structures in the red fluvial sandstones and grey-green lacustrine strata. Approximately 200 m west from the concrete breakwater a lacustrine unit also contains laminites in which fish scales may be seen.

5.14 Volcanic breccia in the volcanic neck at Ness of Duncansby, with Duncansby Head in the background.

Locality 7. Ness of Duncansby [ND 390 739]

Continue along the shore to the western side of the Ness of Duncansby (Fig. 5.12) where a small vent is exposed on the shore; the vent rocks form irregular black reefs near high tide mark. The vent agglomerate includes numerous baked sandstone fragments, but there is little evidence of alteration of the surrounding sandstones. Proceed around the Ness, noting another intercalation of grey-green strata of lacustrine origin in the red fluvial sandstones, and locate exposures of the larger vent. Here, the vent material is a nepheline basalt tuff (Fig. 5.14) which contains crystals of augite and biotite together with fine-grained ultrabasic rocks which might be bombs. Country-rock clasts are dominantly sandstone but also include granites and metamorphic rocks brought up the vent from underlying basement. Dykes and intrusions of nepheline-basalt cut the agglomerate. A K–Ar date of 255 Ma on the nepheline-basalt supports a Permian age for the vent. It is not known if the two exposure areas of vent rocks represent a single vent or two discrete vents.

From here one can return to John o' Groats or continue into Bay of Sannick [397 735] to examine the sandstones on the eastern limb of the gentle syncline that forms the Ness. At Bay of Sannick dip increases to 40° from the usual 10–20°, and in the bay, at low tide, a lacustrine unit with dark grey laminites can be seen, which is possibly the equivalent of the John o' Groats fish bed. If continuing the excursion on foot, walk to the car park by the lighthouse on Duncansby Head; alternatively return to John o' Groats and drive to the lighthouse.

Locality 8. Duncansby Head [ND 405 735]

From the lighthouse car park take the path towards Duncansby Stacks. The cliffs of Duncansby Head are formed of the Mey Subgroup which is faulted against the John o' Groats Sandstone Group. Walk round the ends of the geos (narrow inlets in cliff) and observe the generally thin-bedded nature of the Mey Subgroup and their grey-green colouration. Numerous seabirds nest on the ledges, providing whitewash and a characteristic aroma in spring and summer. Beyond 'The Knee' a path leads to the beach from a metal gate in the fence [ND 403 727]. The path utilises a gully eroded along the fault separating the Mey Subgroup from the John o' Groats Sandstone (Fig. 5.15). On the foreshore at the foot of the gully (low tide required) the fractured rocks of the fault zone can be observed. Most geos are eroded along faults or shatter zones in the rocks that form weaknesses exploited by the erosive power of the sea.

Here, the John o' Groats Sandstone (Fig. 5.16) lacks lacustrine intervals, and high-energy fluvial deposits with trough and planar cross-beds and low angle planar bedding with primary current lineation are seen. Rip-up mud clasts are frequent in channel bases, and beds up to 1 m thick wedge out laterally within 10 m. A broad cyclicity is seen at the southern end of the bay with cross-bedded sandstones on a sharp erosive base grading up into thinner-bedded parallel and ripple laminated sandstones. If the

5.15 Fault gully giving access to locality 8. John o' Groats Sandstone Group on left and thin-bedded flagstones of Mey Subgroup on the right.

5.16 The John o' Groats Sandstone at Locality 8, Duncansby Head.

tide is low it is possible to clamber to Thirle Door, through which there is a good view of the Duncansby Stacks, and also of cycles in the sandstones of the main cliff.

Return to the path out of the bay and walk south along the cliff top for views of the Duncansby Stacks. Lateral variation in the bedding of the sandstones of the stacks can be observed with binoculars. Beds are generally laterally continuous and seldom greater than 1 m thick, possibly indicative of rapid lateral migration of small shallow river channels on the alluvial fan surface.

South Head, Wick
Purpose
Examination of cycles of the Lybster Subgroup of the Lower Flagstone Group. Sedimentary structures are well displayed in the walls of disused coastal quarries.

Access
Follow signs to the Castle of Old Wick from the A99 about 600 m south of the bridge over the Wick River in the centre of Wick. Pass the coastguard station and follow the coastal road to the parking area at the end of the road.

Locality 9. South Head, Wick [ND 373 492]

The cliff has been extensively quarried and the old quarry faces show good detail of the sedimentary structures of the flagstones. Several cycles, typical of the Lybster Subgroup, are exposed between the parking area and the coastguard station. The sloping quarry floors are very slippery when wet and care is required. The flagstone cycles here are dominated by the C Association of Donovan (1980) with thinner sequences of Associations A, B and D (Fig. 1).

In the quarry below the parking area (Fig. 5.17) the cyclicity is picked out well by colour changes. The quarry floor marks the top of a development of D, which is pale grey to green with whitish sandstone beds. This passes up into dolomitic,

5.17 Cyclicity in the Lybster Subgroup at South Head, Wick. Lithological Association D in foreground and at top of quarry face (pale colour). Central part of face consists of grey to black Association C (see Fig. 5.18 and text).

orange-weathering C Association, seen at the base of the quarry face. The grey central part of the quarry face is also C, and water was deepest during deposition of this part of this cycle, but there are no fish bed laminites present. The succession then passes back into orange dolomitic strata and a pale band of D forms the top of the face. Near the entrance to the quarry there are good weathered exposures, below the main quarry floor level where ripples, polygonal desiccation cracks, and loading features can be seen in D. Above the quarry floor are exposures of laminated mudstone and fine sandstone with extensive development of subaqueous shrinkage cracks which are strongly compacted (Fig. 5.18). Brown-weathering dolostone beds up to a few centimetres thick show pull-apart features, and also repetition by low angle thrusts (Fig. 5.18). The dolostones were lithified early in diagenesis and behaved as competent layers, whilst the shales deformed plastically.

The grey, central part of the quarry face shows an excellent section in 'C' Association flagstones with hundreds of mudstone-fine sandstone couplets on a centimetre scale. The base of each sandstone bed is sharp, and sandstone fills underlying lenticular cracks which have been strongly deformed during compaction (Fig. 5.18). The tops of the sandstone beds may be sharp or grade rapidly to mudstone. The sandstone laminae are micaceous and contain pyrite; the dark colour of the mudstone is due to the preservation of organic matter. The origin of the lenticular sand-filled cracks was assigned to subaqueous shrinkage of the mud due to salinity changes (Donovan and Foster, 1972), and rare examples of hopper-shaped crystal pseudomorphs after halite are present indicating that brines were present. Rogers and Astin (1991), and Astin and Rogers (1991) argued that the cracks were subaerial in origin and were filled by wind-blown sand. However, the grain-size, sorting, mineral composition and ripple style support deposition in water. The author concurs with Donovan (1980) and favours a subaqueous origin for the great majority of the lenticular cracks (Trewin, 1992). Futhermore the rocks in question were formed during what is interpreted as the deepest water part of this cycle. In the next quarry to the north at Trinkie swimming pool, a fish-bed laminite (A Association) is exposed in the cliff. The laminite is carbonate-rich, and only rare fish scales have been seen. The laminite grades above and below into thin developments of B Association laminites with a few subaqueous shrinkage cracks. Veins of carbonate and hydrocarbon cut the organic-rich laminites. The transition from D Association of the Trinkie quarry floor to the deep lacustrine laminites takes place very rapidly, showing that there was little sedimentation during lake transgression. A much greater thickness was deposited during the regressive phase of sedimentation as material was eroded from the lake margins and transported towards the lake depocentre.

Proceed north from Trinkie, and in the next quarried area examples of the C and D Associations of Donovan (1980) can be seen with wave ripple-marked surfaces and polygonal desiccation cracks. At this locality, cross-sections of sand-filled cracks which resemble gypsum pseudomorphs can be seen, but many cracks have no resemblance to such pseudomorphs. A great variety of lenticular, polygonal and mixed pattern cracks can be seen at this locality. Horizons already bearing lenticular cracks appear to have been exposed, either causing the lenticular cracks to control polygonal patterns, or to have large desiccation polygons superimposed on the lenticular crack systems.

5.18 A Cut and acid-etched cross-section of typical sand-filled lenticular shrinkage cracks in Association C. **B** Sand/ mud couplets with shrinkage cracks enhanced by weathering in quarry face. **C** Orange weathering dolomitic beds and disruption features near base of quarry section. Lower Flagstone Group, South Head, Wick.

Sarclet

Purpose
To examine the Sarclet Group of the Lower Old Red Sandstone and structural deformation features.

Access
Turn off the A99 in Thrumster at the signpost to Sarclet, and drive to the end of the road, taking a left fork on the way. At the end of the road there is parking room for a few cars on grass, and a small bus could be turned, traffic permitting. Do not block access to the new houses.

Locality 10. Sarclet [ND 351 343]
From the car park area a rocky cliff top can be seen some 500 m to the north, and cliff top exposures of the Sarclet Sandstone Formation can be examined from the north side of the harbour to this point. The sandstones are fully quartz cemented and extensively fractured, possibly due to the proximity of the convergence of the Helmsdale, Great Glen and Wick fault systems immediately offshore at this point. The sandstones are medium-grained and cross-bedding and parallel lamination are present in generally parallel-sided beds up to 50 cm thick.

Several directions of fracturing are present, and the directions vary locally in the Sarclet area. Fracture zones are associated with folds, one prominent syncline plunging to 020°. Other structural features are related to low-angle dislocations overlain by folded and truncated strata. A good example (Fig. 5.19) is exposed in the cliff face below the clifftop exposure visible from the car park. A sheet of disrupted strata rests on an undulating sub-horizontal slide plane which is roughly parallel to the underlying bedding. The cliffs are accessible with care at this point and the dislocation plane can

5.19 Slide plane underlain by relatively undisturbed sandstones and overlain by folded and fractured strata. Cliff top exposure viewed from cliff ledge, locality 10, Sarclet. Further information in text.

be examined. The rocks below are normally bedded and relatively undisturbed, but those above are highly fractured and have numerous quartz veins a few millimetres wide. The rocks were certainly well cemented when the structure developed.

To the north of this locality a fault terminates the exposure of Sarclet Sandstone, and the Riera Geo Mudstone Formation occurs to the north in vertical cliffs which are crowded with nesting seabirds in spring and summer.

Return to the harbour and follow the track down into the bay. The quarried cliff on the north side has hard, parallel-bedded sandstone with internal structures which include wispy lamination defined by thin irregular mud laminae which resemble adhesion ripples, and also sorted laminae of coarse rounded sand grains with grain-size variation on a laminar scale. Deposition was probably affected by aeolian processes.

Viewed from the south side of the bay, large-scale low-angle dislocations are apparent in the cliffs at the seaward end of the north side of the bay. Proceed carefully up a cliff path out of the bay on the south side and walk south along the cliff top. Exposures here show the downward transition to the Sarclet Conglomerate Formation. The Formation comprises alternating conglomerate and pebbly sandstone units. The clasts are up to 30 cm diameter and are composed of granite, schist, quartzite and notably basalt and andesite, indicative of a nearby area of possibly contemporaneous volcanism. The conglomerates are in general poorly sorted, and the clasts mainly subrounded. Deposition was probably by braided streams in an alluvial fan environment. The rocks are extensively fractured, with fractures being more closely spaced in the thinner-bedded units. Return to the starting point along the cliff top path.

ITINERARY 5.3
Old Red Sandstone of Brims Ness to Dunnet Head

Brims Ness
Purpose
To examine the cycles of the Upper Flagstone Group of the Middle ORS.

Access
Turn off the A836 three miles west of Thurso at [ND 057 695] beside a bungalow and follow the single track road (no turning place for bus) towards the large farm building beside the ruined Brims Castle above Port of Brims. Drive through the farmyard, go through a gate to the right to the cliff-top parking area. The section can be examined starting in the small bay in front of the parking area, and heading west along the coast.

Locality 11. Brims Ness [ND 040 715]
The lithologies present are typical of the Upper Flagstone Group. On BGS Sheet 115E the strata at Brims Ness are mapped as Latheron Sub-Group. Recent revision of the stratigraphic nomenclature and fish faunas to the west of the Bridge of Forss Fault (BGS Dounreay sheet (2005) and Newman and den Blaauwen (2007) place these strata in the Crosskirk Bay Formation. The presence of *Dickosteus* indicates that they can be correlated with the Spital Flagstone Formation to the east of the fault (see Fig. 4 in Introduction). The succession comprises cycles averaging 6.5 m

in thickness. Donovan (1980) recognised 14 cycles of which 8 contain carbonate laminite ('fish bed') members, but not all of those contain fish. The cycles represent episodes of lake transgression and regression. In the terminology of Donovan (see Fig. 5.1), DCABCD sequences represent major transgression–regression episodes and DCBCB and DCD sequences less severe transgression and regression. This section was analysed by Hamilton (1986) who by using frequency analysis, and assuming a rate of deposition of 0.28 mm/yr from average varve thickness, produced two prominent peaks corresponding to periods of 22.5 and 88 Kyr. It is implied that climatic cycling associated with Milankovich periodicities was responsible for lake transgression and regression.

On the shore in the bay below the ruined building, flagstones have typical lenticular cracks, here associated with high concentrations of pyrite. Sulphate values in the lake were variable, probably caused by periodic evaporitic concentration of salts (Hamilton and Trewin, 1988). The cliffs on the west side of the inlet show a transition to 'D' facies with numerous wave-rippled sandstone surfaces.

Proceed towards the point and examine the numerous blocks of carbonate laminite which occur at the top of the beach. The rock weathers pale grey but it is dark grey on fresh surfaces. The lamination represents alteration of carbonate (calcite and dolomite) and organic laminae. Microstylolites have developed along some organic laminae and the rock is exceptionally hard and difficult to split. Lenticular chert nodules up to 1 cm thick occur in some fish beds and larger vertically oriented nodules, similar to those described by Parnell (1986) from Orkney, are also present. Fish are present including *Dickosteus thrieplandi*, *Gyroptychius milleri*, *Mesacanthus*, *Diplacanthus*, *Homosteus* and the dipnoan *Pinnalongus saxoni* for which this is the type locality (Newman and den Blaauwen, 2007, and see Loc. 24 below). The scales are bituminised and tend to part from the matrix when exposed. Disruption features in the laminites include bedding parallel, soft-sediment deformation features and later conjugate folds and fractures; the latter frequently containing calcite and black hydrocarbon residues (Fig. 5.20). These residues are the remnants of oil formed

5.20 Deformation features in carbonate laminites of a fish bed at Brims Ness. Coin 27mm.

from the organic-rich laminites during burial. The cyclic nature of the sequence can be examined on the foreshore at mid to low tide.

Continue past the cemetery wall to the start of sea cliffs. Here, the dip flattens into a small faulted syncline flanked by an anticline with the axis trending N–S. The crest of the anticline is faulted and fractured in the upper part of the beach, and the steeply dipping strata form a ridge on the beach; on the lower shore this structure grades into an unfaulted northward-plunging anticline.

Return to Brims Castle and examine the exposures in Brims Harbour. The NNE–SSW Bridge of Forss Fault passes through the harbour and the rocks are extensively shattered over a wide zone by numerous minor faults. Further from the fault, granulation seams are common in the sandstone and nodular cements of dolomite are present. The strata east of the fault are similar to those of the Thurso shore (Loc. 13), and a fish bed exposed 250 m east of the fault contains *Millerosteus minor*, confirming the general correlation.

Holborn Head Quarry
Purpose
To observe features of the sand-poor Upper Flagstone Group lacustrine facies and an associated fish bed rich in *Osteolepis panderi*. This fish bed forms a marker band within the Upper Flagstone Group, and has also been found at Lythmore, Skinnet, and in Cairnfield Quarry (Weydale). This fish bed lies at the top of the Latheron/Spital 'subgroups', and thus modifications to the existing geological maps (BGS sheets 116W and 115E) are required and are in progress using new information on the fish faunas. This illustrates the difficulty in mapping the Caithness flagstones in the poorly exposed interior of Caithness.

Access
From the A836/A9 junction at the western side of Thurso take the A9 to Scrabster. After about 1 km, before reaching the harbour, turn left on the Scrabster Housing access road. Follow the road round to the right and turn left into St Clair Avenue. Park at Scrabster Hall (green building) at the end of the metalled road by the football field. It is a brisk 20 minute walk up the track to the quarry, which is situated on the cliff edge. The track has a left fork just before the quarry, but keep straight on to enter the quarry. Note that this is a cliff-top quarry, and much of the waste was pushed over the cliff edge into the sea. The cliff edge is **dangerous** and should be avoided; the cliff is vertical to overhanging with a drop of about 100 m to the sea. The quarry had been abandoned for many years but working has been resumed on a small scale.

Locality 12. Holburn Head Quarry [ND 080 710]
The floor of the quarry is a bedding surface which dips gently seaward towards the cliff top.

The fish bed lies below the main quarry floor but is exposed in one area near the quarry entrance where numerous fish fragments can be found. The fish bed is a carbonate/organic laminite in which the carbonate is a micritic mix of ferroan dolomite and calcite with a micronodular texture. Bundles of dolomitic laminae occur and are the host laminae for abundant *Osteolepis panderi* (Fig. 5.21) which

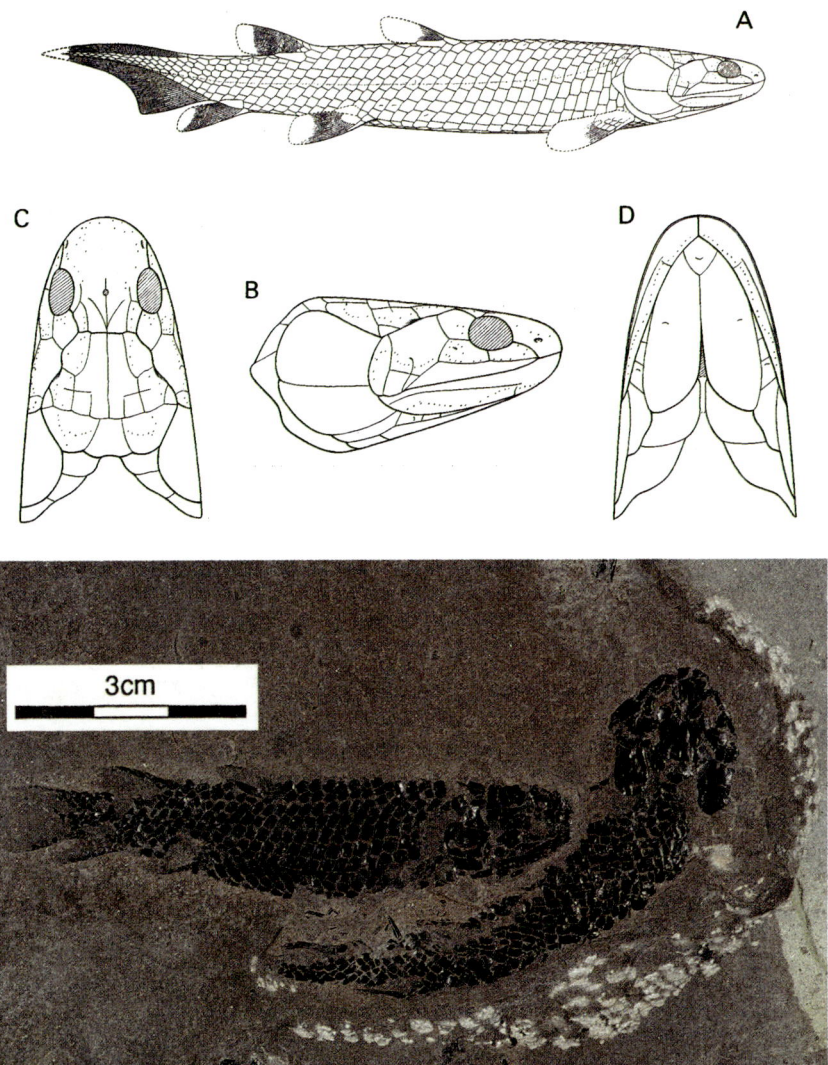

5.21 *Osteolepis panderi*. **A** Reconstruction of lateral view, together with dorsal, lateral and ventral aspects of head (After Jarvik, 1948). **B** Well-preserved, articulated specimens of *O. panderi* from Cairnfield, near Thurso.

died in mass mortalities (Hamilton and Trewin, 1994). The fish carcasses have variable states of disarticulation, with the most highly disarticulated material occurring in the silty upper and lower parts of the bed and the better articulated material in the central, carbonate-rich, part of the bed. The degree of disarticulation increases with increase in depositional energy of the environment, which in turn reflects water depth. Apart from the abundant *Osteolepis panderi*, other fish recorded from this fish bed (various localities) are *Thursius*, *Mesacanthus peachi*, *Dipterus valenciennesi*, *Homosteus* and *Cheiracanthus*.

The strata overlying the fish bed in the quarry comprise 5 m of thin-bedded, dominantly grey-weathering flagstones of lacustrine origin which are laminated on a millimetre scale with micronodular calcareous laminae and silt to fine sand laminae. A few beds of fine sandstone up to 5 cm thick with ripple lamination are present. These are slightly lenticular and erode up to 1 cm into the underlying laminae. The most abundant structures are lenticular sand-filled shrinkage cracks (Fig. 5.22). In the tip material nicely weathered examples of a variety of forms of shrinkage cracks can be found, and there are also examples of small sand-filled crystal pseudomorphs (possibly after gypsum).

5.22 Typical sand-filled lenticular shrinkage cracks from locality 12, Holborn Head Quarry.

Pennyland Shore, Thurso
Purpose
To examine typical cycles of the Mey Subgroup of the Upper Caithness Flagstone Group. This section shows a higher proportion of sandstone and more evidence of subaerial exposure than at Brims Ness. Stratigraphically this section lies in the *Millerosteus* fish zone and is younger than the section at Brims Ness (Fig. 5 in Introduction).

Access
The section is exposed on the shore and in cliffs to the west of Thurso beach, and adjacent to the Camping Site. Cars can be parked on the road beside the beach, but coaches are advised to stop near Thurso Caravan Site, from where the party can follow the cliff-top path and steps to the sandy beach. The eastern part of the section requires low tide. There are several access paths to the section, but take care not to be cut off on a rising tide. The rocks on the wave-cut platform are very slippery when wet.

Locality 13. Pennyland Shore, Thurso [ND 114 687 to 108 692]
The Mey Subgroup cycles exposed in this section show a dominance of Association D of Donovan (1980) when compared with those at Brims Ness. The succession is sandier and has less carbonate. The fish beds have fine lamination restricted to the central parts of the beds and fish (*Millerosteus minor*) are generally disarticulated. Fluctuations in lake depth are represented by rapid transitions from Association A (fish beds) to D. (exposed playa lake), with only weak development of Associations B and C.

At the eastern end of the section, immediately west of the concreted sea defences, thin-bedded grey-green strata of cemented fine sandstone and green shale display excellent examples of both polygonal and lenticular cracks. Both wave and current ripples are present and sandstone beds frequently have internal loading features which form 'pseudonodule' beds. Small-scale trough cross-bedding with sets to 10 cm is also present. The environment was one of frequent exposure with sand transported by currents in ripple sheets, and intermittent shallow water cover in which wave ripples formed.

Between the rock tip with the 'Danger' notice and the concreted cliff steps there are examples of surfaces encrusted by stromatolitic algal sheets and domes. In one case the stromatolites, which weather orange-brown, have colonised a desiccated surface following flooding, and stromatolite growth is clearly related to the desiccation crack morphology.

Below the steps with railings on the landward side of a fault, a grey laminated siltstone fish bed is exposed – but is covered at about half tide. Fish material is disarticulated and widely scattered, probably by current activity, and the typical fine lamination of the fish beds is only poorly developed. The fish bed is cut by thin sandstone dykes, which have compaction features and were thus emplaced prior to the completion of compaction of the fish bed. The strata above the fish bed show a regressive lake sequence, passing up into green sandstone and mudstone with polygonal desiccation cracks.

5.23 A Large sand-filled polygonal desiccation cracks formed due to subaerial exposure. **B** Current ripples formed in shallow water. Locality 13, Pennyland Shore, Thurso.

Continue along the shore, noting the cyclic nature of the sequence. In the second prominent gully past the concreted steps, and on the foreshore, excellent examples of polygonal desiccation cracks are seen in plan and section (Fig. 5.23). In the west wall of the gully a lenticular current-rippled sandstone bed 13 m wide and up to 15 cm thick is exposed in cross section. This is one of several examples of wide, shallow channels with erosive bases, and filled by a single flood event.

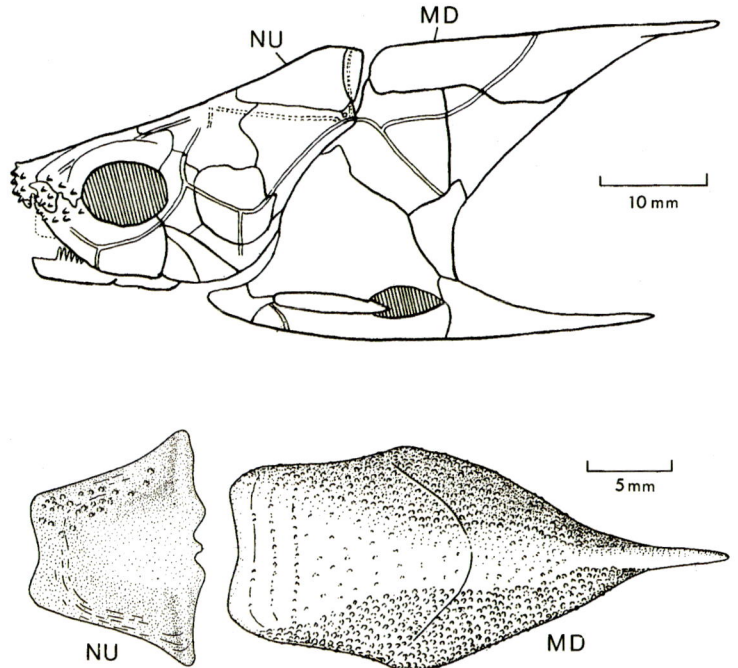

5.24 Reconstruction of lateral view of head and thoracic region of *Millerosteus minor* (Miller) (after A. Desmond).

Near the point, about 100 m east of a low cliff with access path, another fish bed with *Millerosteus* (Fig. 5.24) is exposed. The fish are generally disarticulated, but characteristic plates can be recognised. Sandstone dykes also affect this bed, show only minor compactional features, and are related to joint patterns.

Discrete cross-bedded sandstone units up to 4 m thick occur towards the top of the Mey Subgroup section, and are well exposed in the promontory east of the isolated stack on the seashore. The sandstones are fine- to medium-grained and show a variety of cross-bedding and lamination features (Fig. 5.25). The base of the sandstone includes typical fluvial trough cross-bedding with abundant rip-up clasts of green shale. However, parts of the sandstone consist of low-angle cross-bedding with reactivation features. The sandstone is well sorted and mica-poor, and is finely laminated on a millimetre scale. Granule lags containing well-rounded coarse sand grains occur at the base of some cross-bed sets, and examples of upward-coarsening laminae typical of aeolian ripples are present. Lenticular cross-sections of grain flows are also present. It is probable that the sand bodies were formed by local aeolian reworking of fluvially deposited sands as small dunes and aeolian rippled sandsheets. The sandstones retain some porosity, and there is irregular dark staining by hydrocarbon residues indicating that oil migrated into these sandstones during burial.

At the end of the section the sandstone below the pill box at the west end of a bay with tank traps is similar to that described above with both fluvial and aeolian features present. Pyrite nodules are common in greenish sandstone at this point.

5.25 Cross-bedded sandstones of mixed fluvial and aeolian origin. Promontary near isolated stack below building on cliff top, Thurso shore [ND 111 691].

Dunnet
Purpose
Examination of the Upper ORS Dunnet Sandstone of Dwarwick and Brough, deformation associated with the Brough Fault, and exposures of vent fill.

Access
The most easily accessible localities are situated at the north end of Dunnet Bay near Dwarwick Pier, and on the shore north of the small landing place at Brough opposite Little Clett (Fig. 5.26). Exposures in vertical cliffs can be viewed from the most northerly point of the Scottish Mainland at Dunnet Head. Low tide is required to complete the shore section at Brough, but some exposure at Dwarwick can be seen at all states of the tide. Take the B855 off the A836 in Dunnet Village, the minor road to Dwarwick Pier is not advised for coaches.

Locality 14. Dwarwick Pier [ND 207 713] **Head of Man Bay** [ND 203 719]
From the B855 turn off in Dunnet, continue for about 100 m and carry straight on along the minor road where the B855 bears to the right, and follow signs to Dwarwick Pier where vehicles can be parked.

The bay between Dwarwick Head and Head of Man also contains good exposures of the Dunnet Sandstone, and effects of faulting can be seen. To reach this bay follow the road back from Dwarwick and carry straight on having passed the museum (Mary Ann's Cottage), keep to the left and stop at the drive entrance to the large white house (marked on some maps as 'Northern Gate Ho.', but it is not a

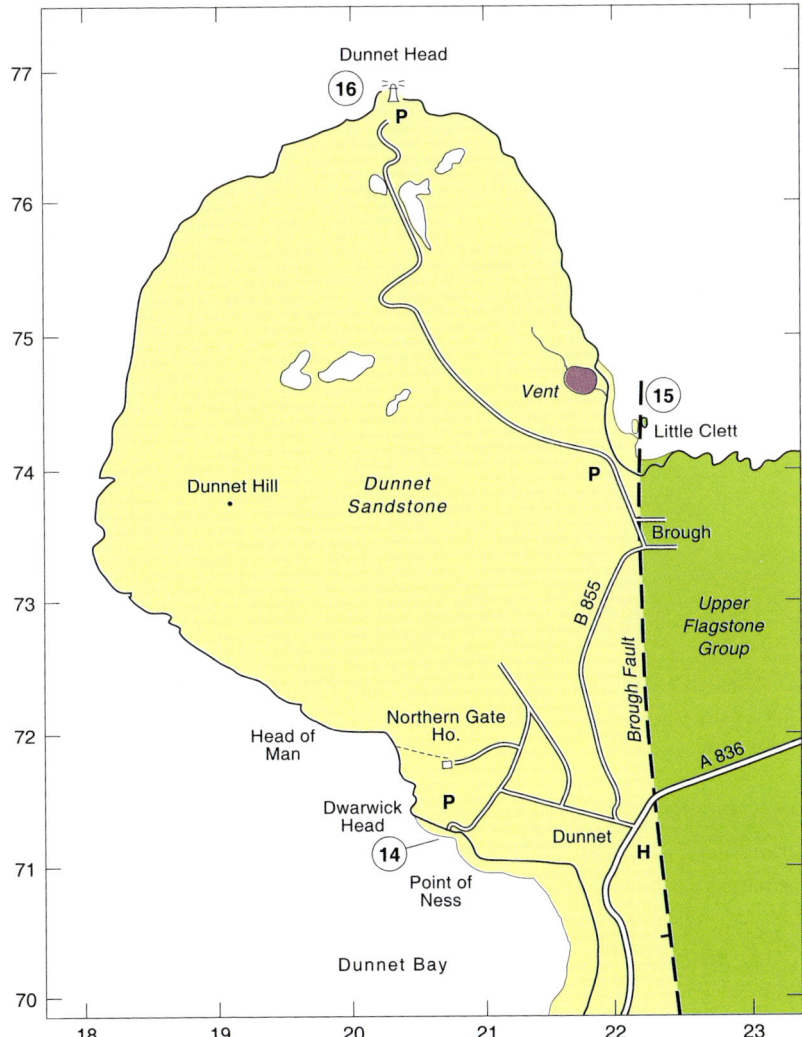

5.26 Locality map for Dunnet Head area, itinerary 5.3.

hotel and parking here is not suitable for coaches). Take the footpath passing to the north of the house which leads down to the beach. Low to half tide is required.

The cliffs NW of Dwarwick Pier (Fig. 5.27) display typical fluvial sandstones of the Upper ORS. Red to buff-coloured, medium- to coarse-grained, quartzose sandstones in beds up to a metre thick are typical. The dominant structure is trough cross-bedding, which frequently has soft sediment deformation features towards the tops of beds. Abundant red and green mudstone clasts, both angular and rounded, are concentrated at the bases of troughs. Low-angle bedding with millimetre-scale lamination displays primary current lineation, and planar cross-bedding is also present. Colour variations in the sandstone represent variable oxidation states of iron and differential erosion accentuates variations in cementation. Cementation is dominantly by quartz overgrowths on quartz grains.

5.27 Cliffs of fluvial cross-bedded red sandstones of the Upper ORS to the northwest of Dwarwick Pier. Loc. 14.

5.28 Soft sediment deformation in cross-bedded fluvial channel sandstones in Upper ORS to the SE of Dwarwick Pier. Loc. 14.

The depositional environment was one of braided fluvial channels, probably on a low-angle alluvial fan. Rapid lateral migration of channels and consequent sediment reworking prevented mudstone preservation except in the form of rip-up clasts of dried mud incorporated into channels. General transport direction was to the NE (McAlpine, 1977). To the SE of the pier, low cliffs and an extensive wave-cut platform permit examination of the cross-bedding and soft-sediment deformation structures in three dimensions, some deformational trough features being over 3 m wide (Fig. 5.28).

If desired the party can also visit the bay to the north between Dwarwick Head and Head of Man. It is a 2 km walk to the bay following the route given above, or a small vehicle can be driven part of the way as described.

Locality 15. Brough, Clett Harbour [ND 221 740]

From Dwarwick Pier return to the B855 and turn left to Brough and Dunnet Head (Fig. 5.26). Stop where the road skirts the cliff top with views of Little Clett, a track not suitable for cars descends to the pier. Low tide is required for this locality.

The most obvious feature is the Brough Fault, which forms a wide fractured zone. The stack of Little Clett consists of extensively fractured ORS. On the foreshore the Upper ORS is vertical due to drag against the fault, but the dip flattens out rapidly towards the cliffs.

The Upper ORS has similar structures to those at Dwarwick, and deposition was dominantly fluvial. If tidal conditions permit, scramble 500 m north along the shore to the cliff break. The cliff exposure north of the small stream consists of about 4 m of quartz-cemented sandstone in two units separated by a pebble bed. This waterlain sandstone overlies a mixed sequence including less well consolidated, low-angle cross-bedded and laminated sandstone. This sandstone appears to be aeolian in origin, consisting of well-sorted sand, with fine lamination and low-angle internal truncations. Mudstone clasts are generally absent. Elsewhere, McAlpine (1977) recognised aeolian dune and sandy playa facies in the Dunnet Sandstone. He concluded that winds dominantly from the south-west reworked the fluvial sands and formed small dunes with interdune playas on the distal areas of the low-angle alluvial fans.

A volcanic vent is exposed in outcrops some 100 m up the Burn of Sinnigoe near this point. Outcrops are poor, but breccias with volcanic fragments and clasts of ORS and basement lithologies can be seen, often in a sandy matrix. As Peach observed (*in* Crampton and Carruthers, 1914) it appears that the vent pierced the Upper ORS prior to its lithification. Contemporaneous Upper ORS volcanics occur at the base of the Hoy Sandstone in north Hoy. However, the volcanic vent at Duncansby (Loc. 7) is considered to be of Permian age.

Locality 16. Dunnet Head [ND 202 768]

It is worth taking the opportunity to visit Dunnet Head for excellent views of inaccessible cliffs of the Dunnet Sandstone, which display the generally good lateral continuity of the bedded units. McAlpine (1977) recognised nine lithostratigraphic units that could be correlated between Dunnet Head and the island of Hoy, which can be seen to the north.

ITINERARY 5.4

Old Red Sandstone basin margin deposits at Red Point, Port Skerra, Baligill and Sandside Bay

General purpose

To demonstrate the nature of the sub-Devonian unconformity and basin marginal deposits of the Orcadian Basin.

General access

All localities are close to the A836 between 13 and 18 miles by road to the west of Thurso. For those staying at Helmsdale it is a fine scenic drive along the A897 (mainly single track with passing places) up Strath Helmsdale and down Strath Halladale to this area. This road passes Baile an Or, site of the Helmsdale gold rush, and subject of Excursion 6. Access to individual localities is given with the locality details.

Red Point

Purpose

To demonstrate the unconformity between basement and the ORS, which includes an example of a coincident lake margin deposit with limestone deposited on a dipping basement surface.

Access

About a mile to the west of Reay on the A836 there is a lay-by with an Information Board at [NC 932 646] on the north side of the road (Fig. 5.29). The lay-by is big

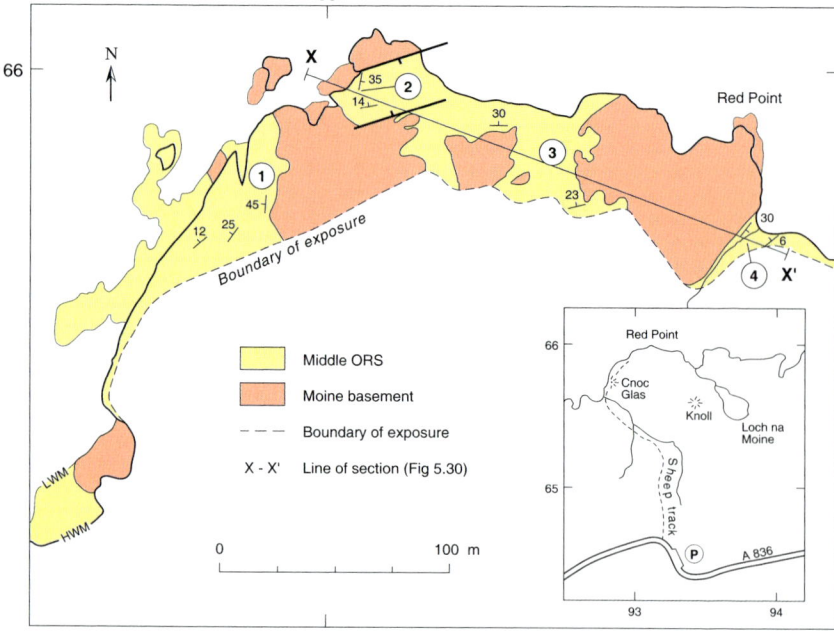

5.29 Locality map of the Red Point area, locality 17 (Modified from Donovan 1975).

enough to park a coach. From the lay-by there are views to the east of Dounreay across Sandside Bay. There are no well-defined paths in this area, and the walk is not advised in poor visibility without a compass. Beware of boggy areas in old peat workings. From the lay-by walk west on the road for 150 m, then head north across the moor along the side of the ridge with a valley to the east. Head towards a grassy knoll (Cnoc Glas) on the coast, which will come into view. The stream, sometimes boggy, can be crossed near the edge of the sea cliff. Follow the ill-defined coastal path about 200 m to the east to Point 1 on Figure 5.29. This walk takes about 30 minutes. A total time of at least 2 hours is required for this excursion.

The cliffs in this area are dangerous and must be treated with caution, particularly when wet.

Locality 17. Red Point [NC 930 659 to 933 659]

Locate yourself with respect to the geology as shown on Figure 5.29. A lot of time can be spent examining the details of these outcrops. The following notes cover some of the main points; refer to Donovan (1975, 1978) and Janaway and Parnell (1989) for further information and discussion. Donovan demonstrated that a relief of 30 m+ exists on the basement surface in this locality area (Fig. 5.30).

Point 1. This striking exposure (Fig. 5.31) shows steeply (45°) dipping strata resting on gneissose and granitic basement. Above a thin calcite cemented breccia, grey limestones mantle the surface and display folding due to downslope movement. Massive limestones grade laterally downslope into laminites indicating increasing water depth down the flank of the basement high. This is an example of a lacustrine limestone deposited directly on basement during a period of lake highstand, a coincident lake margin in the terminology of Donovan (1975). The limestone shows irregular silicification in places and Janaway and Parnell (1989) recognised a lower dolomitic unit within the limestone. The limestone is truncated by a coarse breccia of basement and limestone clasts in a sandstone matrix. The limestone clasts are up to 30 cm long and represent reworking of the underlying limestone that was clearly fully lithified at the time of erosion and transport.

Point 2. Here a prominent ridge of red breccia with calcitic cement lies between two areas of basement with faulted margins. The sandstones with breccia tongues (Fig. 5.32) beneath the breccia show original dips that rapidly flatten out. Imbrication of clasts can be seen in the breccia. Wave ripples and desiccation cracks are present, indicating periods of exposure, but some breccia beds are draped by thin carbonate-rich laminites deposited in deeper water. The breccias appear to be lake margin deposits, perhaps representing a beach facies.

Point 3. In this area the relief on the basement can be observed, and the sandstones and breccias overlying the unconformity examined in more detail. Sandstones are thin-bedded with wavy and subparallel lamination and contain lenses of breccia resting in shallow hollows with erosive bases. On the east side of this broad gully some large boulders on the unconformity surface are set in a laminated sandstone matrix that was washed into place between the boulders. Some areas overlying the uncomformity are dominated by sandstone and others by breccia. Clamber out of the east side of the gully and follow the cliff to the next valley and descend the

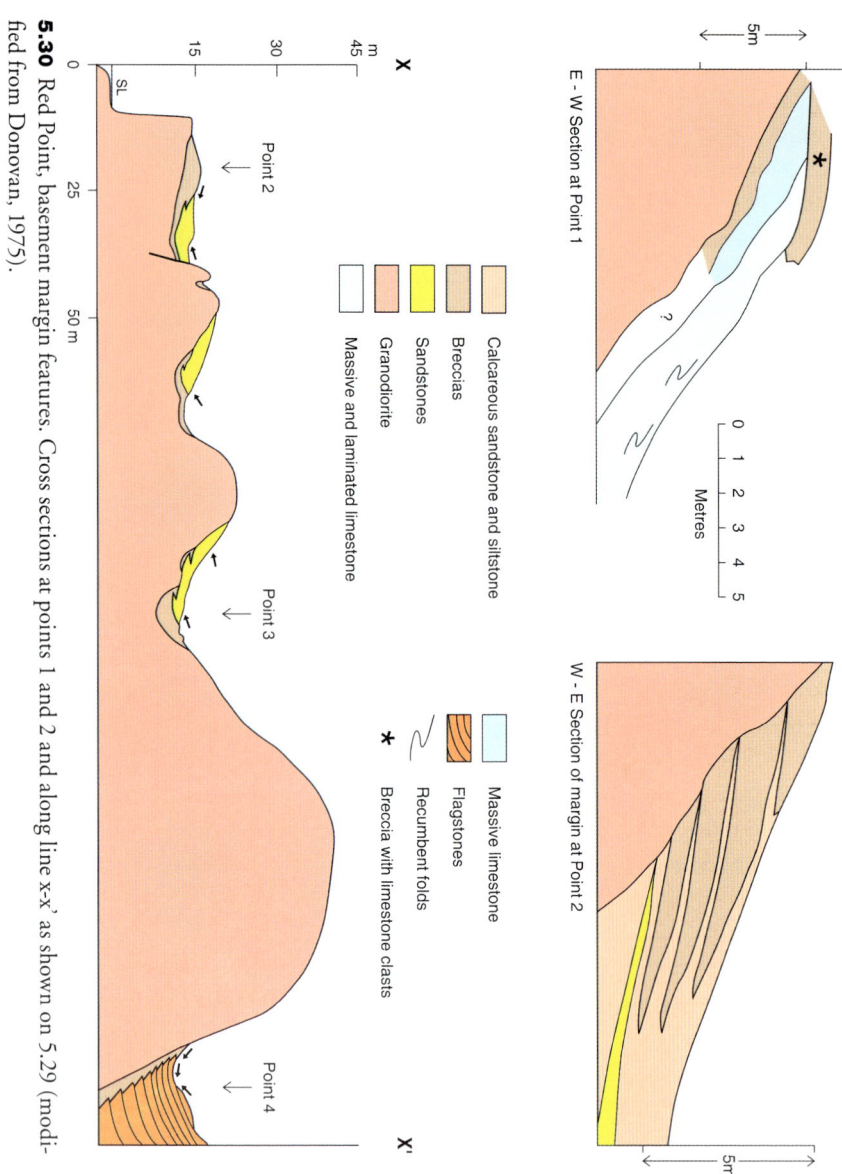

5.30 Red Point, basement margin features. Cross sections at points 1 and 2 and along line x-x' as shown on 5.29 (modified from Donovan, 1975).

grassy slope by the stream at the head of the gully to Point 4 (Fig. 5.33). The steep grassy slopes are very slippery when wet.

Point 4. The junction between a steeply dipping surface of gneiss and flagstone facies is exposed in this valley. The western wall of the valley exposes gneiss with granite veins, and flagstones make up the eastern side. Although minor faulting is present, as would be expected given the competency difference between the gneiss and flagstones, a lateral transition can be seen from gneiss outcrop, though 5 m of breccia containing gneiss and limestone clasts, to flagstones with scattered angular gneiss clasts. The flagstones display typical shallow-water current and wave ripples, laminated sandstones with primary current lineation, and also horizons with polygonal mud cracks. The limestone clasts in the breccia resemble the limestone seen at

5.31 Exposure at Point 1, Locality 17, Red Point. Steeply dipping limestone mantles the basement and is overlain by breccia.

5.32 Exposure at Point 2, Locality 17, Red Point. Rapid lateral transition from marginal breccia downslope into lacustrine flagstones.

Point 1 and probably originated from erosion of a limestone that was deposited on the gneiss during a highstand of the lake, and subsequently eroded when it became exposed as water level dropped.

Close examination of the surface of the gneiss outcrop reveals fissures with relics of laminated and massive limestone, and carbonate-cemented patches of breccia

5.33 Gully at Point 4, Locality 17, Red Point. View to north of steep exhumed margin of basement hill of gneiss cut by granite veins at left of gully, and lacustrine flagstones in valley floor and on right.

still plastered to the gneiss surface. At times of low lake level the gneiss hill was coincident with the playa lake margin. The coarser deposits seen at Point 3 on the other side of the gneiss outcrop appear to represent a lower stratigraphic level.

Return along the cliff-top to Cnoc Glas and retrace the route to the lay-by.

Port Skerra
Purpose
To examine the features associated with the unconformity between Moine basement and the Old Red Sandstone.

Access
Turn off the A836 at the Melvich Hotel (Fig. 5.34) into Portskerra village. Take the right-hand fork at the road junction and drive to the last row of cottages where a track bears right down to a concreted slipway at [NC 878 663]. A minibus can be parked near the cottages, or down the track at the slipway. There is only room for a couple of vehicles at the slipway, so it is advisable to check that space is available before driving down to the slipway. Large coaches should be left at the Melvich Hotel.

Locality 18. Port Skerra [NC 878 663]
From the slipway the major feature of the geology is clearly seen in the cliffs and reefs protecting the bay (Fig. 5.35). An irregular surface of Moine gneisses is overlain unconformably by conglomerates and sandstones of the Old Red Sandstone, here believed to be Middle ORS; but direct evidence is lacking.

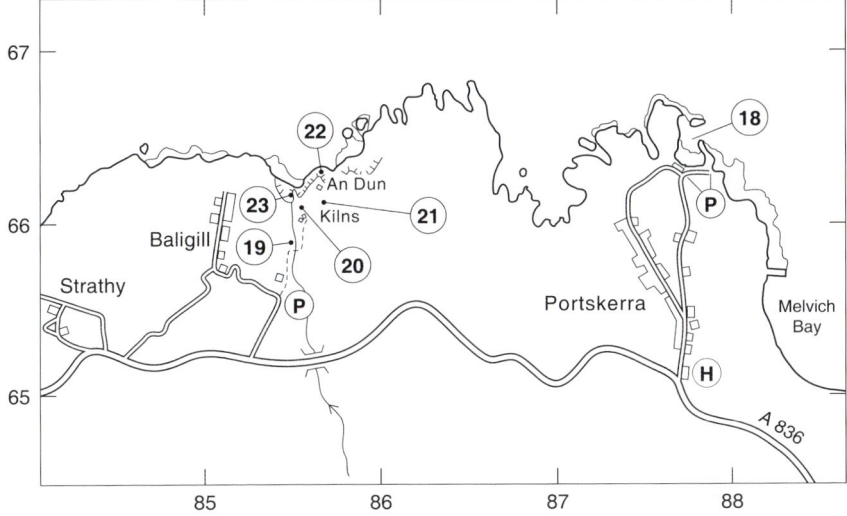

5.34 Locality map for Port Skerra and Baligill, localities 18 to 23.

5.35 View to the west of Portskerra Bay from the track to the slipway. Knolls of Moine gneiss are draped by Old Red Sandstone.

In the area of the slipway, and to the south-east, coarse breccias with a green sandy matrix rest on grey- to red-banded gneiss (Fig. 5.36). The gneiss is extensively migmatised and more than one melt phase is present, the melt layers being commonly deformed into complex folds. This multiphase high-grade metamorphism and deformation took place in the Ordovician Grampian Orogeny. These gneisses, together with amphibolites representing metamorphosed basic dykes, have been extensively intruded by undeformed late pink granite veins that are probably connected with the Strath Halladale Granite, which is dated at $c.425$ Ma. Fracturing is locally intense in the basement, but the ORS is less fractured, indicating that some fracture sets predate ORS deposition and were opened by Devonian weathering (Fig. 5.37).

5.36 Banded and folded Moine gneiss in reef at the end of the slipway, Port Skerra.

5.37 Unconformable contact between intensely jointed Moine gneiss and locally-derived Old Red Sandstone breccia. Near end of slipway, Port Skerra.

In the bay, outcrops on the beach are dominated by hard silica-cemented sandstone with crude parallel lamination in beds up to 1 m thick. Most beds contain scattered angular clasts of basement gneiss and granite which occur in a floating texture in the sandstone matrix. Deposition was rapid with little reworking and the transport distance short. These are probably amalgamated deposits of flash floods.

At the end of the bay the sandstones can be examined (half to low tide) where they abut the tops of knolls of the basement gneiss. The sandstone beds show little modification adjacent to the basement knolls, which have only a thin (20 cm) veneer of locally derived breccia. In the hollows between knolls a greater thickness of breccia is usually present. The first basement knoll contains a sediment-filled fissure over 3 m deep and up to 20 cm wide; part of the fissure fill forms the seaward face of the knoll. A sandstone bed exposed near the low tide mark some 20 m from the basement knoll shows excellent convolute lamination with an amplitude of 60 cm. Current and wave ripples are also present in the sandstones. The sandstones forming the gentle synclinal structure between two basement knolls are mainly parallel laminated, but some cross-bedding and minor channelling is present. The synclinal feature appears to be caused by differential compaction over the irregular basement surface. At low tide it is possible to scramble up the second knoll and examine further details of the unconformity, and also a large mass of diorite cut by sheets of undeformed granite within the Moine basement.

If time permits a visit can be made to the small cove to the west of the headland where the unconformity can also be seen at [NC 8755 6633]. Further details, particularly of the basement, are included by Strachan *et al.* (2009c in press) in Exc. 13 of the revised excursion guide to the Moine.

Baligill
Purpose
Demonstration of marginal features of the Orcadian Basin.

Access
Coaches should drop people at the minor road junction at [NC 853 652] (Fig. 5.34). Smaller vehicles can park 400 m down the road near the corner, where a track descends towards the sea past a cottage. Take care not to block farm gates. Binoculars are useful for observing the geology of the cliffs in this area. Walk down the track past the cottage towards localities 19–23.

Locality 19. Balligill Limestone Member [NC 855 659]
Follow the track to the bridge over the stream. Downstream of the bridge (*c.*20 m) on the west bank a grey carbonate laminite fish bed is exposed, which is directly overlain by a brown medium-grained sandstone with ripple lamination. This is a typical basin margin feature, there being a transition from deep lake to very shallow conditions without the usual transitional facies. This sandstone fines up into thin-bedded flags and a second carbonate laminite is seen in the outcrop with a wall built on top. These laminites are part of the Baligill Limestone Member of the Bighouse Formation (British Geological Survey, 2005). Both these laminite beds contain fish including *Pinnalongus saxoni*, *Coccosteus cuspidatus*, *Mesacanthus* and *Cheiracanthus* (record in Newman and Dean, 2005). It is now considered that the characteristic Achanarras fish bed fauna is not seen in the area west of the Forss Fault. The Balligill Limestone Member lies within the Lower Caithness Flagstone Group, about 270 m below the probable equivalent of the Achanarras fauna.

Locality 20. Lime kilns [NC 855 661]

Continue on the track on the east side of the stream through a metal gate to the old lime kilns. At the base of the quarried face opposite the kilns (Fig. 5.38) a laminated limestone is exposed in which laminae have stylolitised contacts along organic-rich laminae, and large stylolites cut the bedding at a high angle. Above these lacustrine laminites is a sequence of green shales and dominantly fine-grained sandstones which have excellent loading structures on the bases of beds 5–50 cm thick. A few scattered granite pebbles are also present. Grading is present in some beds and partial Bouma sequences (bcde) deposited by waning currents are seen. This part of the sequence appears to have been deposited in moderately deep water by density currents, but the occasional granite pebbles show that basement was exposed nearby. This sequence is overlain, apparently conformably, by a sandstone with pebbles of quartz and granite at the base. This sandstone is dominantly medium-grained and is cross-bedded with sets up to 1 m in amplitude. At the top of the outcrop the sandstone is thin-bedded and contains current ripples and small-scale cross-bedding. The sandstone is a shallow water deposit of fluvial or lacustrine delta origin and was deposited as a result of coarse sediment prograding over the deeper water deposits during a period of low lake level. Sandstones exposed to the west of the outcrop show cross-bedding that is probably of aeolian origin, suggesting retreat of the lake and establishment of subaerial conditions similar to those seen in the Fresgoe Sandstone Member at Sandside Bay (Loc. 24 below).

5.38 Section at Locality 20, by the lime kilns. Shallowing-up section from lacustrine laminite at base of cliff to fluvial/ aeolian sandstones at top.

5.39 Angular clasts of basement gneiss in limestone that drapes the gneiss surface. Loc. 21, near An Dun.

Locality 21. [NC 857661]
Continue on the path to the east of the outcrop beside a fence. Basement is seen exposed on a ridge to the left. Cross the ridge to a valley with a small burn. On the west side of the valley there is an exposure of limestone with angular fragments of gneiss, and dipping towards the burn (Fig. 5.39). A small pile of quarried limestone blocks of similar lithology lies nearby. Cross-bedded sandstones are exposed on the other side of the valley. It can be seen that the gneiss formed a hill that was mantled by limestone before later burial by sandstones. The scattered limestone outcrops in this area were quarried for lime production for agricultural use. Cross the burn and ascend the path to the promontary of An Dun overlooking the sea 100 m to the north.

Locality 22. An Dun [NC 857662]
Only a few stones of the dun (fort) remain. The high cliffs have a steep grassy slope on the east side and limestone can be seen dipping steeply off the mound of basement below, although at the top of the cliff strata are nearly flat-lying (Fig. 5.40). This locality is dangerous and not suitable for closer examination, particularly in poor conditions. Walk 200 m round the cliff top to the next headland to the east and look back to the cliff below An Dun. A steeply dipping surface of grey limestone mantles the basement gneiss surface. At a prominent ledge on the cliff face the reddish gneiss is exposed and can be viewed from a distance (Fig. 5.41). Similar rock-types are more safely seen at Red Point.

Locality 23. [NC 855 662]
Follow the cliff-top path back westwards towards the bay where a fault-bounded ridge of sandstone can be observed with binoculars. Northward-directed cross-

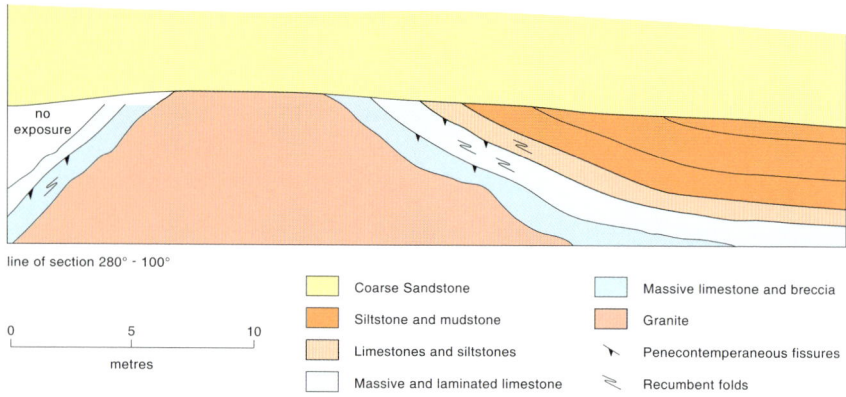

no
exposure
line of section 280° - 100°
0
5
10
metres
Coarse Sandstone
Siltstone and mudstone
Limestones and siltstones
Massive and laminated limestone
Massive limestone and breccia
Granite
Penecontemperaneous fissures
Recumbent folds

5.40 East-west section through the basement knoll at An Dun. Loc. 22 (modified from Donovan, 1975).

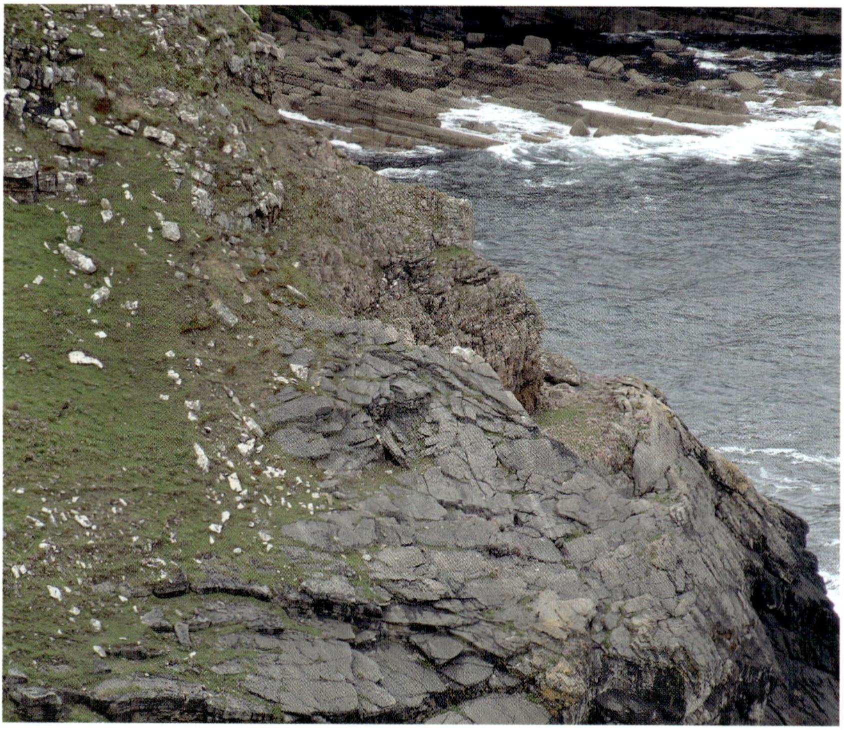

5.41 View of the cliff face below An Dun showing outcrop of gneiss beneath grey lacustrine limestone that drapes the steep gneiss surface.

bedding in the cliff comprises sets up to 1 m thick. Style varies from low-angle planar cross beds to units in which the cross-bedding is asymptotic to the base of the set. Reactivation features can be seen in several sets. These sandstones appear to be deposits of bars in broad, shallow river channels flowing north into the Orcadian Basin. In hand specimens well-rounded quartz grains can be seen and aeolian reworking is implied. Return up the track to the vehicle.

Note: In the first edition of this guide Baligill Quarry [NC 852 657] was included and the fish fauna taken to indicate the Achanarras faunal level. Some of the fish determinations on which this conclusion was based have been shown to have been incorrect following further collecting in the area by Newman and den Blaauwen (see British Geological Survey, 2005). The quarry is now badly overgrown.

Sandside Bay
Purpose
To examine part of the flagstone sequence to the west of the Bridge of Forss Fault where the Achanarras Limestone member and its characteristic fauna are not present. An interesting unit (Fresgoe Sandstone Member) of aeolian sandstones occurs within the dominantly lacustrine successsion. The stratigraphy of the area is described on BGS special sheet for Dounreay (British Geological Survey, 2005)

Access
Turn off the A836 to Sandside Bay in Reay village (Fig. 5.42). The road passes a few houses and after the cattle grid there is parking space on the right, but it is easier to park at the public conveniences a few hundred yards further on. There is also a larger parking area at the end of the public road near Sandside harbour, where a coach could be turned. Warning notices are displayed regarding radioactive particles

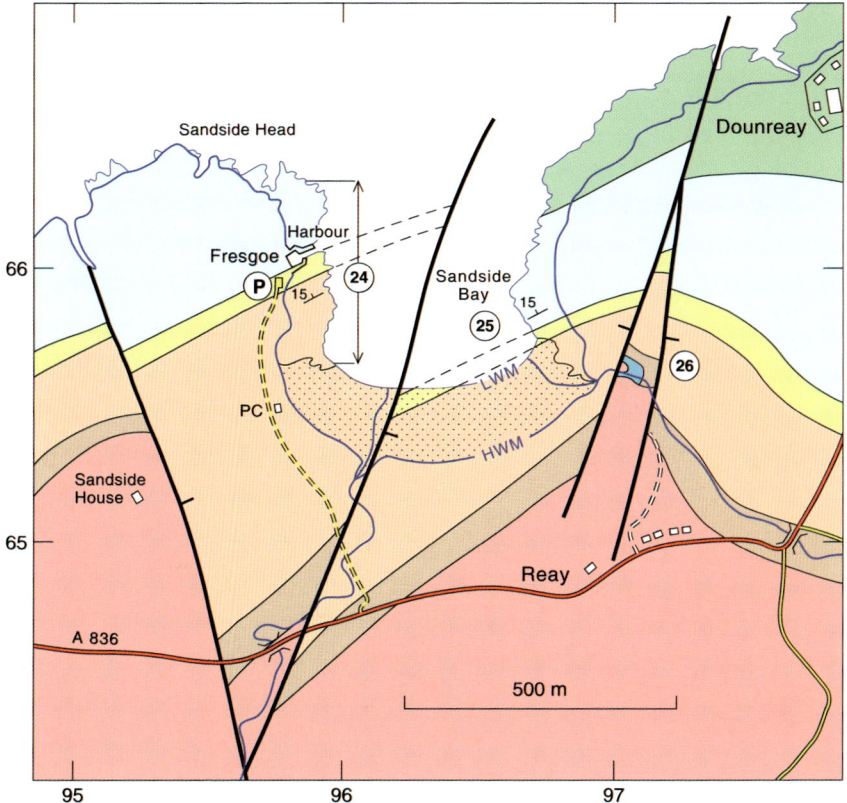

5.42 Locality map, Sandside Bay.

that have been washed onto the beach from Dounreay. Regular monitoring of the beach for such material takes place.

Locality 24. Sandside Bay, western side [NC 958 655 to 958 662]

The head of the bay is a sandy shore lacking rock exposure, but good exposures occur on both the western and eastern sides of the bay, and display a similar succession with the Fresgoe Sandstone Member forming a prominent feature. The first exposures encountered on the shore are part of the Bighouse Formation. This cyclic sequence contains fish bed laminites in which *Thursius macrolepidotus*, *Pinnalongus saxoni*, *Coccosteus cuspidatus*, *Diplacanthus*, *Mesacanthus* and *Cheiracanthus* have been recorded (British Geological Survey, 2005). *Pinnalongus* is a dipnoan (lungfish) described by Newman and Den Blaauwen (2007). It is present in the Lower Flagstone Group to the east of the Bridge of Forss Fault, but ranges up into the Upper Flagstone Group west of the fault. The main depositional thickness of the cycles was deposited in shallow water close to the lake margin, and wave and current ripples and polygonal desiccation cracks are common (Fig. 5.43). Sandstone beds dominate the sequence and are frequently loaded into the green shale to produce 'ball and pillow' structures.

Continue north to the first low cliff at the top of the beach, where a rapid transition from aeolian sandstone to deep lake laminite can be seen (Fig. 5.44). The prominent sandstone at the base of the cliff comprises well-sorted fine sandstone with typical aeolian cross-bedding. The sand dunes must have invaded the dried lake floor during a period of low lake level. The marginal position of Sandside Bay with respect to the edge of the basin is demonstrated by basement outcrops of diorite about 500 m to the SE of the bay; thus the sand dunes probably fringed the basin margin at times of low lake level. The top of the aeolian sandstone was planed off and reworked as the lake waters rose, and some 80 cm of rippled sandstone and shale deposited in shallow water. As the lake deepened further, the shale content increases, the colour turns to grey, and small sand-filled subaqueous cracks dominate the next 50 cm. There follows 50 cm of typical lacustrine laminite with fish remains deposited in the deepest lake phase. Thus the deepening phase of the lake resulted in little sedimentation.

The cyclic sequence continues to the north with a prominent fish bed present in front of the house. At the northern end of the house beneath the wall the base of the Fresgoe Sandstone Member is exposed. This sandstone is exposed on the foreshore and in low cliffs from here to the Harbour wall. This exposure lies adjacent to the parking area at the end of the public road. Seats and an information board can be seen at the cliff edge. The sandstone is about 30 m thick, and displays spectacular aeolian cross-bedding on a large scale with cross-bed sets to more than 2.5 m in thickness, and continuous for tens of metres laterally (Fig. 5.45). The sandstone is generally pale yellow, fine-grained and well sorted. Transport direction of the dunes was generally to the north.

Near the harbour wall there is an area of disruption of the dune bedding with folding and steep to vertical dips in the cross-bed lamination. Parts of the sandstone are homogenised, with lamination becoming indistinct. It is probable that a major flood caused disruption and collapse of the dunes to produce these structures.

5.43 Rippled sandstone overlying polygonal desiccation cracks. Bighouse Formation, Sandside Bay.

5.44 Cliff exposure with bed of aeolian sandstone followed by rapid transition to laminated fish bed. Bighouse Formation, Sandside Bay.

5.45 A Cross-bedded aeolian sandstone of the Fresgoe Sandstone Member near Sandside Harbour wall. **B** Lacustrine flagstones overlying truncated top of the aeolian Fresgoe Sandstone. East side of Sandside Bay, Loc. 25.

This sandstone can also be examined on the eastern side of the bay (Loc. 25) where the disruption features seem to be absent, but the top and base of the sandstone are better exposed. The top is remarkable for the flat truncation surface (Fig. 5.45) with only minor reworking by water, below lacustrine flagstones. It is possible that as the lake and the water table rose, any topography that the dunes retained was removed by wind deflation rather than reworking by water.

Leave the beach and walk around the harbour, ascend the stone steps at the north end of the building, and return to the shore north of Sandside Harbour. From here to the northern limit of the beach the Sandside Bay Formation can be examined. This formation marks a return to cyclic flagstone deposition similar to the Bighouse Formation but with the addition of sandstone beds deposited by relatively high-energy flows (Fig. 5.46). The beds occur in several cycles and are best exposed in the cliff opposite a stack at the northerly limit of the beach; a concreted pipe is present in the gully leading north beside the cliff. The sandstone beds range from 5 to 50 cm in thickness and have erosive bases, with a few of the thicker beds displaying flute moulds. Internally the beds are dominated by parallel lamination, which shows primary current lineation on split surfaces; some climbing ripple lamination is also present. The tops of several beds are current rippled. Beds can be traced laterally for tens of metres with little variation. One prominent bed has been disrupted by loading. These beds were clearly deposited by short-lived flash-flood events, and sheet-floods seem a likely mechanism, since channels are not

5.46 Section north of Sandside harbour with sandstone beds deposited by flash floods.

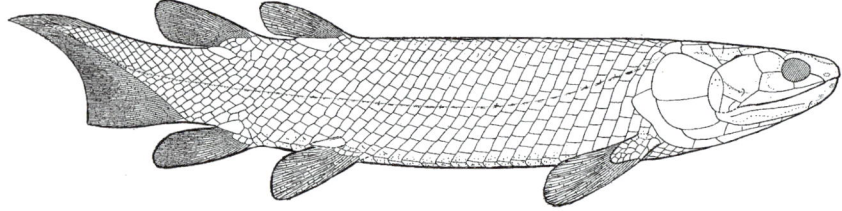

5.47 *Thursius macrolepidotus*. Reconstruction (after Jarvik, 1948) and specimen from Sandside Bay.

developed. In this location close to the basin margin infrequent storms probably resulted in rapid run-off with the flood spreading as a sheet over the flat exposed playa surface. Current directions are generally northerly, but with a wide range between east and west, inplying that the lake bed was flat and floods arrived at the site from several local sources.

A few metres above these sandstone beds the succession becomes thin-bedded; wave ripples and polygonal desiccation cracks are present. A rapid lake transgression took place resulting in fish-bed laminites. The laminites can be examined 40 m to the north in the wall of the gully. This fish bed contains *Thursius macrolepidotus* (Fig. 5.47) and acanthodian fragments.

Locality 25. Sandside Bay, eastern side [NC 968 658]

If desired the succession described above can also be examined on the east side of the bay, and compared with that on the west. The aeolian sandstones of the Fresgoe Sandstone Member are well exposed and the top of the member is clearly seen (Fig. 5.45). The fish bed laminites of the cycles can also be correlated across the bay (British Geological Survey, 2005). Take care not to enter the area of the Dounreay Nuclear Site; there is extensive security control in this area.

Locality 26. Burn of Isauld [NC 970 657]

Exposures in the Burn of Isauld can be found by following the burn inland for about 100 m from the eastern corner of Sandside Bay. Small outcrops in and near the stream display breccias, limestone and basement diorite. The limestone, termed the Aryleive Limestone Bed (British Geological Survey, 2005) is up to 10 m of massive to laminated limestone that rests unconformably on basement diorite. It is mapped as being locally present at the base of the Portskerra Conglomerate Member, which in part overlies the limestone. The environmental interpretation is comparable to the situation seen at Red Point (Loc. 17) and at An Dun (Loc. 22). It does not seem possible to prove the relative ages of these limestones, but they might have formed during the same deep lake phase, draping drowned basement highs where there was no source of clastics. They are unlikely to occur in basement lows where sandstones and breccias accumulated. Eventually clastics covered the basement highs as sedimentation increased at the basin margin, and the next deep lake events seen in the area are the fish beds of the Bighouse Formation seen at locality 25, and Baligill (Loc. 19).

Excursion 6

Kildonan gold

C. M. Rice

Purpose
To pan for gold at the site of the 1868–69 Gold Rush, to examine the Moine country rocks and especially the granite and quartz veins from which the gold may have been ultimately derived, and also, the glacial deposits which are the immediate source of the gold in the burns.

Access
If possible the excursion should begin in the Timespan Heritage Centre, beside the old bridge in Helmsdale. Timespan is open from Easter to October. Mon–Sat 10–5, Sun 12–5. At the centre the 'Goldrush Tour' can be hired. This is a hand-held audiovisual GPS that will take you on a guided tour through the history of the goldrush. Gold panning equipment can be purchased or hired in Helmsdale at Strath Ullie Crafts by the harbour.

Leave Helmsdale by the A897 to Forsinard. The journey to Kildonan is about 9 miles and since it is a single track road about 20 minutes should be allowed (excluding stops). The minimum time to complete the excursion is half a day.

Permits to pan for gold are no longer required, but visitors are asked to place details of their visit in a box at the information point at Baille an Or. Follow any instructions given in the information display. Panning is allowed only between the bridge at Baille an Or [NC 912 214] and the wooden bridge at [NC 917 228] (Fig. 6.1). Allow about 45 minutes to walk from the Baille an Or to the wooden bridge. Digging into the banks is not permitted as it results in serious erosion. Other areas on the Estate may be visited for geological studies but the Factor must be informed of the intended route, especially during the shooting season (normally mid-August to mid-October).

Maps useful to this excursion are O.S. 1:25000 Sheet NC 82/92 and Geological Survey 1:63360 Sheet 109.

Advice on panning
Equipment should include gold pan, sieve (about 10 mm), shovel, bottle, horse-shoe magnet, tweezers, long-handled narrow-headed spoon for crevicing, and wellingtons.

Select sites where the current velocity is significantly reduced, e.g., insides of bends, reduced gradient, around large boulders and natural riffles. Dig as deep as possible and attempt to reach bedrock. Potholes, especially beneath waterfalls, and rock crevices may also trap gold particles. Very fine gold can often be obtained by panning the sediment trapped by moss growing on the stream bed. Select areas of coarse sediment, i.e. about pebble size. This is because gold is a very dense material

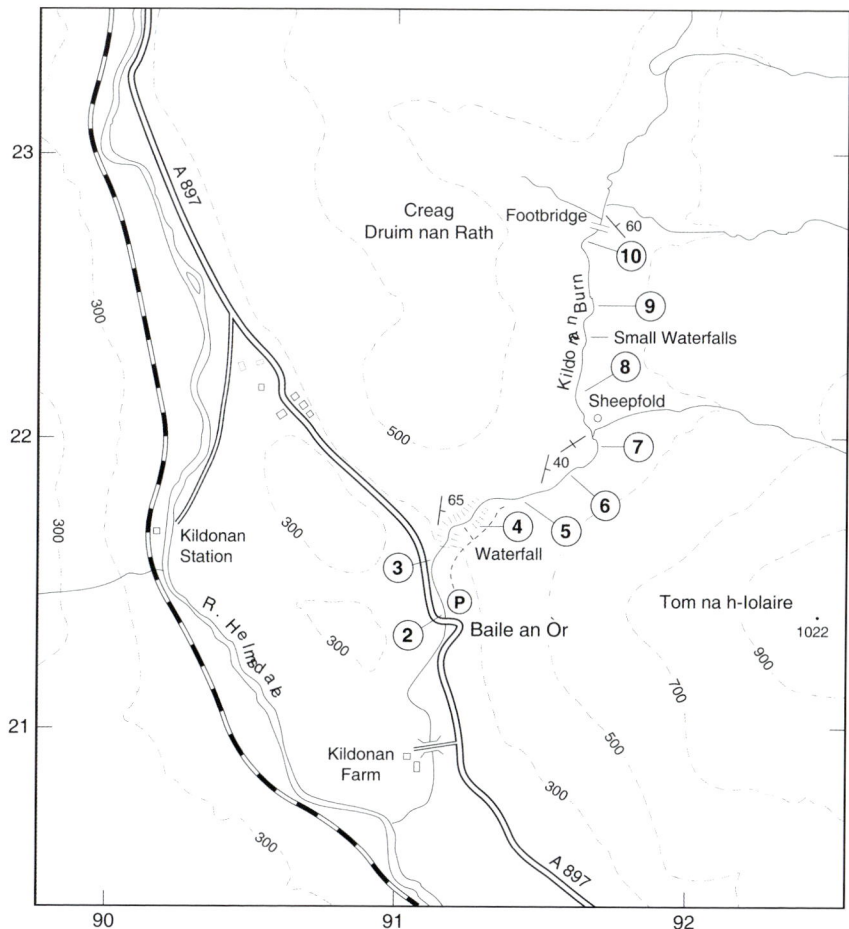

6.1 Locality map of Kildonan Burn area.

(about 8 times that of quartz) and is hydraulically equivalent to much larger but less dense particles.

Panning efficiency is a matter of skill and practice! Firstly remove the large material by hand or by sieving, but wash the coarse material thoroughly before discarding it from the pan. Use a swirling motion so that all material (sediment and water) in the pan is in motion. This will allow any gold and other heavy minerals to fall to the bottom. The light material is allowed to wash over the rim of the pan. A good guide as to the efficiency of your technique is the nature of the material remaining in your pan. At Kildonan it should consist of black iron oxides (magnetite and ilmenite), purple garnet and hopefully, small flakes of gold (Fig. 6.2). False hope may be generated by the presence of minor iron pyrite (fools' gold!) and bronze-yellow flakes of biotite mica. The vast majority of the gold likely to be found is less than 1 mm in size. A lot of people try their luck at Kildonan, so you are likely to be panning material that has been looked at already; if you find nothing at one spot either move on or dig deeper.

6.2 Typical flakes of alluvial gold up to 3 mm in size, and panned from gravel in the Kildonan Burn at Baille an Or.

Introduction

Alluvial gold, in the form of a nugget weighing more than half an ounce, was discovered in the Kildonan burn around 1840. However, it was not until 1868 that gold was found in quantity by a local man (R.N. Gilchrist) who had returned from the Australian goldfields. From 1868 to 1869 at least 3500 oz of gold (valued at about £2 million at present prices (£550/oz)) was obtained from various burns, but the richest deposits were found in the Kildonan and Suisgill burns. This is a very small alluvial deposit by world standards; the Californian goldfields produced 42 million oz. The largest nugget came from the Suisgill burn and weighed 2 oz 17 grams. In 1870 the diggings were closed by the Duke of Sutherland following complaints from sporting and farming interests; the mud washed into the Helmsdale River was ruining the fishing.

Those interested in the historical aspects of the gold at Kildonan should read the booklet on *The Kildonan Gold Rush* by Jack Saxon. Contemporary reports of the gold diggings include Joass (1869); a long article (Anon, 1869a) in the *Illustrated London News* entitled "Something from 'the diggins' [sic!] in Sutherland", and another (Anon, 1869b) in the same publication entitled 'The gold-fields of Sutherlandshire'. The latter article has two fine engravings of the diggings (Figs. 6.3, 6.4). The general geology of the area is described in Read (1931).

Kildonan gold 173

6.3 Engraving of Baille an Or at the time of the 1869 gold rush. Reproduced from *The Illustrated London News*, May 29, 1869.

6.4 Engraving of gold diggers working at Kildonnan in 1869. Reproduced from *The Illustrated London News*, May 29, 1869.

The source of the gold has inevitably attracted considerable interest and the ultimate source is probably granite and quartz veins cutting the Moine metasediments. Auriferous granite pebbles have been found (Joass, 1869) and gold has been panned from deeply altered (probably weathered) granite in Suisgill Burn [NC 904 269] (Dawson and Gallagher, 1965). The granite is considered to belong to the Strath Halladale migmatite complex, which contains both Older (late Precambrian) and Newer Granites (Pankhurst and Sutherland, 1982). Thus, there is considerable uncertainty over the age of the primary mineralisation. A further problem is that the distribution of gold seems to be unrelated to either the intensity of migmatisation or known Newer Granite activity (cf. the nearby Helmsdale Granite). Systematic variations in the composition of the gold from localities in the area indicate a centre of mineralisation in the watershed of the Suisgill burn. Thus, the gold in Suisgill burn is more silver-rich than in the adjacent Kildonan and Kinbrace burns. Recently, an epithermal source has been proposed on the basis of the silver and tellurium content of the gold and the presence of inclusions of polymetallic Bi-sulphides in the gold (Chapman, 2007). Given the close proximity of the Devonian–basement unconformity, this might suggest a Devonian age for the mineralisation.

Subsequent weathering, erosion and both glacial and fluvial transport processes have affected the distribution and concentrations of the gold. The immediate source of the gold is morainic terraces bordering the streams, and especially the lowest part of these (Joass, 1869). Gold grades broadly increase according to the extent of reworking, from which it follows that the primary granite source may have been of low grade rather than a 'Bonanza' deposit. Strongly altered granite occurs in the upper reaches of the gold burns, especially Suisgill, and is probably caused by deep pre-glacial Tertiary weathering. This would have released gold grains from a relatively large volume of granite for natural beneficiation. Further upgrading would have occurred by reworking and concentration during melting and retreat of the Pleistocene ice sheet(s), and by recent fluvial action (Plant and Coleman, 1972), thus the present alluvial concentration represents the results of a multistage process. General information about the glacial history of the Scottish Highlands can be found in Boulton *et al.* (2002).

All of the key elements of the gold story outlined above, with the exception of the deeply altered granite, can be seen in this excursion.

Locality 1. [NC 026 155]
A visit to the Timespan Heritage Centre opposite the Bridge Hotel in Helmsdale is recommended. Here, apart from numerous interesting exhibits on the history and wildlife of the area, one exhibit is devoted to the Gold Rush and shows alluvial gold from the diggings as well as contemporary pictures and panning equipment.

Proceed up the Strath of Kildonan on the A897 towards locality 2. The Strath is a broad flat-bottomed glacial valley devoted to farming and salmon fishing. Grouse and deer are hunted in the adjacent hills. The hillsides are blanketed with drift and exposure is generally very poor. Cross sections of moraine can be examined at Kilphedir (by the bridge [NC 989 186] and about 100 m west of Torrish Burn [NC 968 187]. Generally the drift is unstratified and consists of angular to rounded, mainly psammitic clasts of varying sizes set in a sand-clay matrix. At Torrish Burn a moraine contains a lens of finely bedded sand surrounded and overlain by more typical unstratified drift. The sandy lens represents local fluvial reworking of morainic material. At Torrish Lodge a fine example of a tree-covered esker can be seen on the south side of the road, and in the distance on the southern side of the Strath there is a fine display of hummocky drift-covered ground.

Locality 2. Baile an Or [NC 912 214]

Many miners lived at Baile an Or (the town of gold) during the Gold Rush (Fig. 6.3). Good examples of the country rocks, mainly Moine psammites, can be examined both by and below the bridge. These are pale brown flaggy siliceous rocks with thin pelite partings striking 333° and dipping at 45° to the east. In the area covered by this excursion the Moine rocks show little variation. They are predominently eastward dipping, grey, flaggy to medium-bedded psammites, with occasional thin pelitic bands, which have attained amphibolite facies metamorphism. Exposures are largely restricted to the edge of the burn. On the west side of the bridge the psammites are cut by veins (10 cm scale) of red leucogranite. Both are cut by thin (1 cm scale) quartz veins. Veins of both types are possible primary sources of gold, but do not attempt to find any. The chances of success are remote and it would cause unacceptable damage to the outcrop.

Walk upstream to locality 2, where gravel on both sides of the burn may be panned. Note two terraces on the east side of the burn. Gold has been obtained from the lower portion of such terraces, especially above bedrock.

Locality 3. [NC 911 216]

Here the burn makes a sharp turn and a large waterfall can be seen at the foot of a gorge. The marked change in gradient makes the gravels between localities 2 and 3 attractive targets. On the inside of the bend the burn cuts into and exposes the lower terrace, which is an unstratified mixture of rounded clasts set in a sandy matrix. The gravel bank at this point is a good place to pan.

To reach locality 4 climb the hill above locality 3 to a narrow path along the upper terrace. The path along the lower terrace leads to the waterfall and is slippery and dangerous. This path forks above the waterfall; take the upper path and then descend to the burn above the waterfall.

Locality 4. [NC 913 217]

Pan the large gravel bank on the east side of the burn. Bedrock is close to the surface here. The psammites strike 172° and dip 65° to the east. Retrace your steps up the side of the valley to the path and proceed to locality 5.

Locality 5. [NC 914 218]

The downstream end of this locality is marked by a small 2 m waterfall. At this point the psammites strike across the burn, forming a series of natural riffles. Sediment caught between the riffles can be panned. Look for sediment-filled crevices in the bedrock here. At the waterfall a good example of a massive quartz vein occurs striking across the burn. The psammites strike at 172° and dip about 30° to the east.

About 50 m upstream of the waterfall a small gravel bank close to the bedrock on the east side of the stream can be panned. At this point the burn is roughly parallel to the strike of the psammites. From locality 5 it is possible to walk beside the burn for the remaining part of the excursion.

Locality 6. [NC 916 219]
Here, the valley opens out above the gorge and a terrace is present. The psammites (strike 183°, dip 40° to the east) strike across the burn again, forming a series of riffles. From here to the sheepfold [917 221] gravel banks on both sides of the burn may be panned.

Locality 7. [NC 917 220]
The dip of the psammites varies from about 65° to vertical, probably due to small-scale folding. On the east side of the stream a small antiform occurs with an axial planar cleavage and apparently a sub-horizontal fold axis. The fold geometry is rather obscure due to faulting.

Locality 8. [NC 916 222]
About 90 m upstream of the sheepfold a small S fold crosses the burn. The fold axes plunge steeply to the north. From this point to locality 9 the burn runs through a fairly straight open valley with poor exposure and not many gravel banks are exposed at medium water levels in the burn.

Locality 9. [NC 916 225]
Here, the burn has cut into a terrace and exposed a 50 m section of closely fractured and disturbed psammites, pelites and thin granite and quartz veins, overlain by drift. A large gravel bank on the east side of the burn is worth panning.

Fracturing is most intense in the central part of the exposure where two main directions can be recognised by displacements on a thin 5 cm granite vein crossing the face just above eye level. Possibly these fractures are axial planar cleavages to a conjugate fold. The structure cannot be fully resolved, due to the rock face being partially obscured by downwashing drift.

The structures seen at localities 7–9 suggest that the overall structure of the psammites is probably more complex than a set of eastward dipping beds.

Locality 10. [NC 916 227]
A large gravel bank on the inside of the bend may be panned. On the opposite bank the burn has cut into a terrace, exposing sandy drift. Psammites exposed about 40 m below the bridge are striking at 327° and dipping 60° to the east. Upstream from the wooden bridge two terraces can be easily distinguished.

References

ALLISON, I., MAY, F. and STRACHAN, R.A. (eds) 1988. *An excursion guide to the Moine geology of the Scottish Highlands.* Scottish Academic Press.

ANDREWS, S. 2008. Climatic cyclicity recorded in the Middle Old Red Sandstone of the Orcadian Basin. Univ. Aberdeen PhD thesis (unpubl.).

ANDREWS, I.J. and BROWN, S. 1987. Stratigraphic evolution of the Jurassic, Moray Firth. *In*: Brooks, J. and Glennie, K.W. (eds) *Petroleum Geology of North-West Europe.* Proceedings 3rd Conf. Pet. Geol. N.W. Europe. 785–95. Graham and Trotman.

ANDREWS, I.J., LONG, D., RICHARDS, P.C., THOMSON, A.R., BROWN, S., CHESHER, J.A. and McCORMAC, M. 1990. *United Kingdom offshore regional report: the Geology of the Moray Firth.* London: HMSO for the British Geological Survey.

ANON. 1869a. Something from 'the diggins' in Sutherland. *The Illustrated London News.* April 20, 1–31. 1869b. The gold-fields of Sutherlandshire. *The Illustrated London News.* May 29, 535–8.

ARMSTRONG, M., DONOVAN, R.N. and MYKURA, W. 1978. Western Moray Firth and Caithness. *In*: Friend, P.F. and Williams, B.P.J. (eds) *A field guide to selected outcrop areas of the Devonian of Scotland, the Welsh Borderland and South Wales.* 32–37. Palaeontological Association International Symposium on the Devonian System 1978.

ASTIN, T.R. 1985. The palaeogeography of the Middle Devonian Lower Eday Sandstone, Orkney. *Scottish Journal of Geology* **21**, 353–75.

ASTIN, T.R. 1990. The Devonian lacustrine sediments of Orkney, Scotland; implications for climate cyclicity, basin structure, and maturation history. *Journal of the Geological Society, London* **147**, 141–57.

ASTIN, T.R. and ROGERS, D.A. 1991. 'Subaqueous shrinkage cracks' in the Devonian of Scotland reinterpreted. *Journal of Sedimentary Petrology* **61**, 850–9.

BAILEY, E.B. AND WEIR, J. 1932. Submarine faulting in Kimmeridgian Times: East Sutherland. *Transactions of the Royal Society of Edinburgh* **57**, 429–67.

BARRON, H.F. 1986. Dinoflagellate cyst biostratigraphy and palynofacies analysis of the Upper Jurassic strata at Helmsdale, N.E. Sutherland. Univ. Aberdeen M.Sc. Thesis (unpubl.).

BARRON, H.F. 1989. Dinoflagellate cyst biostratigraphy and palaeoenvironments of the Upper Jurassic (Kimmeridgian to basal Portlandian) of the Helmsdale region, east Sutherland, Scotland. *In*: Batten, D.J. (ed.) *Studies in NW European micropalaeontology and palynology.* 192–213. British Micropalaeontological Society Series, Ellis Horwood, Chichester.

BATTEN, D.J., TREWIN, N.H. and TUDHOPE, A.W. 1986. The Triassic–Jurassic junction at Golspie, Inner Moray Firth Basin. *Scottish Journal of Geology* **22**, 85–98.

BELL, B. R. and WILLIAMSON, I. T. 2002. Tertiary igneous activity. *In*: Trewin, N.H. (ed.) *The Geology of Scotland* (4th Edn), 371–407. The Geological Society, London.

BENTON, M.J. and WALKER, A.D. 1985. Palaeoecology, taphonomy, and dating of Permo-Triassic reptiles from Elgin, north-east Scotland. *Palaeontology* **28**, 207–34.

BERRIDGE, N.G. 1967. Brora Coalfield. Geological report on the evidence of 5 boreholes drilled on November, 1966. I.G.S. internal report.

BERRIDGE, N.G. and IVIMEY-COOK, H.C. 1967. The geology of a Geological Survey borehole at Lossiemouth, Morayshire. *Bulletin Geological Survey G.B.* **27**, 155–69.

BEVERIDGE, R., BROWN, S., GALLAGHER, M.J. and MERRITT, J.W. 1991. Economic Geology. *In*: Craig, G.Y. (ed.) *Geology of Scotland* (3rd Edn). 545–95. The Geological Society, London.

BLAKE, J.F. 1902. On a remarkable inlier among the Jurassic rocks of Sutherland and its bearing on the origin of the breccia-beds. *Quarterly Journal of the Geological Society of London* **57**, 290–312.

BOULTON, G.S., PEACOCK, J.D. and SUTHERLAND, D.G. 2002. Quaternary. *In*: Trewin, N. H. (ed.) *The Geology of Scotland* (4th Edn) 409–430. The Geological Society, London.

BRITISH GEOLOGICAL SURVEY. 2005. Dounreay, Scotland, parts of sheets NC96, ND06 and ND07 Bedrock. 1:25 000 Geology Series. British Geological Survey, Keyworth.

BROOKFIELD, M.E. 1976. The age of the Allt na Cuile Sandstones (Upper Jurassic, Sutherland). *Scottish Journal of Geology* **12**, 181–6.

BROWN, P.E. 1991. Caledonian and earlier magmatism. *In*: Craig, G.Y. (ed.) *Geology of Scotland* (3rd Edn) 229–93. The Geological Society, London.

CARROLL, S. 1990. Terrestrial, fluvial and marginal lacustrine ecosystems in the Old Red Sandstone of the Orcadian Basin. University of Aberdeen PhD thesis (unpubl.).

CHAPMAN, R. 2007. An overview of gold mineralization in the Caledonides of Great Britain and Ireland: insights from placer gold geochemistry. *In*: C.J. Andrew *et al.* (eds) *Digging Deeper* Proc. 9th Biennial SGA Meeting. Dublin. 943–46.

CHARLESWORTH, J.G. 1956. The late glacial history of the Highlands and Islands of Scotland. *Transactions of the Royal Society of Edinburgh* **62**, 769–928.

CHESHER, J.A. and LAWSON, D. 1983. *The geology of the Moray Firth*. Report of the Institute of Geological Sciences, No. 83/5, 32pp.

CLARK, N. D. L. 1999. The Elgin Marvel. *Open University Geological Society Journal* **20**, 16–18.

COLLINS, A.G. and DONOVAN, R.M. 1977. The age of two Old Red Sandstone sequences in southern Caithness. *Scottish Journal of Geology* **13**, 53–57.

COWARD, M.P. and ENFIELD, M.R. 1987. The structure of West Orkney and adjacent basins. *In*: Brooks, J. and Glennie, K.W. (eds) *Petroleum geology of northwest Europe*. Proceedings 3rd. Conference on the Petroleum Geology of NW Europe. 687–96. Graham & Trotman.

COWARD, M.P., FRANCIS, P.W., GRAHAM, R.H., MYERS, J.A. and WATSON, J. 1969. Remnants of an early sedimentary assemblage in the Lewisian complex of the Outer Hebrides. *Proceedings of the Geologists' Association* **80**, 387–408.

CRAMPTON, C.B. and CARRUTHERS, R.G. 1914. *The geology of Caithness*. Memoirs of the Geological Survey, Scotland.

CROWELL, J.C. 1961. Depositional structures from the Jurassic boulder beds, East Sutherland. *Transactions of the Edinburgh Geological Society* **18**, 202–20.

CUNNINGHAM, H.R.J. 1841. Geognostical Account of the County of Sutherland. *Transactions of the Highland and Agricultural Society of Scotland* New Series 7, 73–114.

DAWSON, J. and GALLACHER, M.J. 1965. Alluvial gold in Scotland. *Mining Journal*. March 12. 193.

DEC, T. 1992. Textural characteristics and interpretation of second cycle, debris-flow-dominated alluvial fans (Devonian of Northern Scotland). *Sedimentary Geology* 77, 269–296.

DESMOND, A.J. 1974. On the coccosteid arthrodire *Millerosteus minor*. *Zoological Journal of the Linnean Society* **54**, 277–298.

DONOVAN, R.N. 1973. Basin margin deposits of the Middle Old Red Sandstone at Dirlot, Caithness. *Scottish Journal of Geology* **9**, 203–211.

DONOVAN, R.N. 1975. Devonian lacustrine limestones at the margin of the Orcadian Basin, Scotland. *Journal of the Geological Society, London* **131**, 489–510.

DONOVAN, R.N. 1978. The Middle O.R.S. of the Orcadian Basin. *In*: Friend, P.F. and Williams, B.P.J. (eds). *A field guide to selected outcrop areas of the Devonian of Scotland, the Welsh Borderland and South Wales.* 37–53. Palaeontological Association International Symposium on the Devonian System 1978.

DONOVAN, R. N. 1980. Lacustrine cycles, fish ecology and stratigraphic zonation in the Middle Devonian of Caithness. *Scottish Journal of Geology* **16**, 35–50.

DONOVAN, R.N. and FOSTER, R.J. 1972. Subaqueous shrinkage cracks from the Caithness Flagstone Series (Middle Devonian) of north-east Scotland. *Journal of Sedimentary Petrology* **42**, 309–17.

DONOVAN, R.M., FOSTER, R.J. and WESTOLL, T.S. 1974. A stratigraphical revision of the Old Red Sandstone of North East Caithness, *Transactions of the Royal Society of Edinburgh* **69**, 167–201.

DUNCAN, A.D. and HAMILTON, R.F.M. 1988. Palaeolimnology and organic geochemistry of the Middle Devonian in the Orcadian Basin. *In*: Fleet, A.J., Kelts, K. and Talbot, M.R. (eds) *Lacustrine Petroleum Source Rocks*. Geological Society, London, Special Publications **40**, 173-201.

DUNCAN, W. I. and BUXTON, N, W, K. 1995. New evidence for evaporitic Middle Devonian lacustrine sediments with hydrocarbon source potential on the East Shetland Platform, North Sea. *Journal of the Geological Society, London* **152**, 251–8.

EWING, C.J.C. 1956. *Geological Report on the Brora Coalfield*. N.C.B. Report.

EWING, C.J.C. 1958. *Brora Coal Field – Geological Report on borings*. N.C.B. Report.

FLINN, D. 1992. The history of the Walls Boundary fault, Shetland: the northward continuation of the Great Glen fault from Scotland. *Journal of the Geological Society, London* **149**, 721–6.

FROSTICK, L., REID, I., JARVIS, J. and EARDLEY, H. 1988. Triassic sediments of the Inner Moray Firth, Scotland; early rift deposits. *Journal of the Geological Society, London* **145**, 235–48.

GILLEN, C. 1987. Huntly, Elgin and Lossiemouth. *In*: Trewin, N.H., Kneller, B.C. and Gillen C. (eds) *Excursion guide to the geology of the Aberdeen Area.* 149–60. Scottish Academic Press.

GLENNIE, K.W. 1998. *Petroleum geology of the North Sea*. 4th Edn. Blackwell, Oxford.

GLENNIE, K.W. 2002. Permian and Triassic. *In*: Trewin, N H (ed.) *The Geology of Scotland* (4th Edn), 301–21. The Geological Society, London.

GLENNIE, K.W. and BULLER, A.T. 1983. The Permian Weissliegend of N.W. Europe: the partial deformation of aeolian dune sands caused by the Zechstein transgression. *Sedimentary Geology* 35, 43–81.

GLUYAS, J.G. and HITCHENS, H.M. (eds) 2003. *United Kingdom Oil and Gas Fields Commemorative Millennium Volume*. Geological Society, London, Memoirs 20.

HAMILTON, R.F.M. 1986. Comparative palaeolimnology of the Middle Devonian Orcadian Basin. Univ. Aberdeen PhD thesis (Unpubl.).

HAMILTON, R.F.M. and TREWIN, N.H. 1985. *Excursion to the Devonian of Caithness*. Petroleum Exploration Society of G.B. Aberdeen Branch 1–35.

HAMILTON, R.F.M. and TREWIN, N.H. 1988. Environmental controls on fish faunas of the Middle Devonian Orcadian Basin. *In*: McMillan, N.J., Embry, A.F. and Glass, D.J. (eds) *Devonian of the World*. Canadian Society of Petroleum Geologists, Memoir 14, Vol III, 589–600.

HAMILTON, R.F.M. and TREWIN, N.H. 1994. Taphonomy of fish beds from the Upper Flagstone Group of the Middle Old Red Sandstone, Caithness. *Scottish Journal of Geology* 30, 175–81.

HARKER, S.D. 2002. Cretaceous. *In*: Trewin, N H (ed.) *The Geology of Scotland* (4th Edn), 351–60. The Geological Society, London.

HARRIS, A.L. and JOHNSON, M.R.W. 1991. Moine. *In*: Craig, G.Y. (ed.) *Geology of Scotland* (3rd Edn). 87–123. The Geological Society, London.

HARRIS, T.M. and REST, J.A. 1966. The flora of the Brora coal. *Geological Magazine* 103, 101–9.

HUDSON, J.D. 1962. The Great Estuarine Series (Middle Jurassic) of the Inner Hebrides. Unpublished Ph.D. thesis, Univ. Cambridge.

HUDSON, J.D. 1964. The petrology of the Great Estuarine Series, and the Jurassic palaeogeography of NE Scotland. *Proceedings of the Geologists' Association* 75, 499–527.

HUDSON, J. D. and TREWIN, N.H. 2002. Jurassic. *In*: Trewin, N. H. (ed.) *The Geology of Scotland* (4th Edn), 323–50. The Geological Society, London.

HURST, A. 1980. The diagenesis of Jurassic rocks of the Moray Firth N.E. Scotland. Univ. Reading Ph.D. Thesis (unpubl.).

HURST, A. 1981. Mid Jurassic stratigraphy and facies at Brora, Sutherland. *Scottish Journal of Geology* 17, 169–77.

HURST, A. 1982. The clay mineralogy of Jurassic shales from Brora, NE Scotland. *In*: van Olphen, H. and Veniale, F. (eds) *International Clay Conference 1981*. Developments in Sedimentology. 35, 677–84, Elsevier.

HURST, A. 1985. The implications of clay mineralogy to palaeoclimate and provenance during the Jurassic in N.E. Scotland. *Scottish Journal of Geology* 21, 143–60.

JANAWAY, T.M. and PARNELL, J. 1989. Carbonate production within the Orcadian Basin, Northern Scotland: a petrographic and geochemical study. *Palaeogeography, Palaeoclimatology, Palaeoecology* 70, 89–105.

JANVIER, P. and NEWMAN, M.J. 2005.On *Cephalaspis magnifica* Traquair, 1893, from the Middle Devonian of Scotland, and the relationships of the last osteostracans. *Transactions of the Royal Society of Edinburgh: Earth Sciences* 95, 511–25.

JARVIK, E. 1948. On the morphology and taxonomy of the Middle Devonian Osteolepid fishes of Scotland. *Kungliga Svenska Vetenskapsakademiens Handlingar* 25, pp301.

JOASS, J.M. 1869. Notes on the Sutherland gold field. *Quarterley Journal of the Geological Society of London* 25, 314–26.

JOHNSTONE, G.S. and MYKURA, W. 1989. The Northern Highlands of Scotland. British Regional Geology (4th Edition). British Geological Survey, HMSO, London.

JONK, R. 2003. The origin and diagenesis of intruded sandstones. Univ. Aberdeen PhD thesis (unpubl.).

JUDD, J.W. 1873. The secondary rocks of Scotland. *Quarterly Journal of the Geological Society of London* 29, 97–195.

KELLY, S.B. 1992. Milankovitch cyclicity recorded fromDevonian non-marine sediments. *Terra Nova* 4, 578–84.

KINNY, P.D., FRIEND, C.R.L., STRACHAN, R.A., WATT, G.R. and BURNS, I.M.1999. U–Pb geochronology of regional migmatites, East Sutherland, Scotland: evidence for crustal melting during the Caledonian orogeny. *Journal of the Geological Society, London* 156, 1143–52.

KNELLER, B.C. 1987. A geological history of north-east Scotland. *In*: Trewin, N.H., Kneller, B.C. and Gillen, C (eds) *Excursion guide to the geology of the Aberdeen area*. 1–50. Scottish Academic Press.

KNOX, R. W, O'B. 2002. Tertiary sedimentation. *In*: Trewin, N.H. (ed.) *The Geology of Scotland* (4th Edn), 361–370. The Geological Society, London.

LAM, K. and PORTER, R. 1977. The distribution of palynomorphs in the Jurassic rocks of the Brora Outlier, N. E. Scotland. *Journal of the Geological Society, London* **134**, 45–55.

LEE, G.W. 1925. Mesozoic rocks of East Sutherland and Ross. *In*: Read, H.H., Ross, G. and Phemister, J. *The geology of the country around Golspie, Sutherland*. Memoirs of the Geological Survey 65–115.

LEEDER, M.R., BOLDY, S.R., RAISWELL, R. and CAMERON, R. 1990. The Carboniferous of the Outer Moray Firth Basin, quadrants 14 and 15, Central North Sea. *Marine and Petroleum Geology* 7, 29–37.

LINSLEY, P.N. 1972. The stratigraphy and sedimentology of the Kimmeridgian deposits of Sutherland, Scotland. Univ. Lond. Ph.D. Thesis (unpubl.).

LINSLEY, P.N., POTTER, H.C., McNAB, G. and RACHER, D. 1980. The Beatrice field, Inner Moray Firth, U.K. North Sea. *In*: Halbouty, M.T. (ed.) *Giant oil and gas fields of the decade 1968–1978*. Memoir American Asssociation of Petroleum Geologists **30**, 117–29.

McALPINE, A. 1977. The Upper Old Red Sandstone deposits of Hoy and Dunnet Head, Northern Scotland. University of Newcastle-on-Tyne PhD thesis. (unpubl.).

MACDONALD, A.C. 1985. Kimmeridgian and Volgian fault-margin sedimentation in the northern North Sea area. Univ. Strathclyde Ph.D. Thesis (unpubl.).

MACDONALD, A.C. and TREWIN, N.H. 1993. The Upper Jurassic of the Helmsdale Area. In. Trewin, N. H. and Hurst, A. (eds) Excursion Guide to the Geology of East Sutherland and Caithness. Scottish Academic Press, Edinburgh, 75-114.

MACDONALD, R. and FETTES, D.J. 2007. The tectonomagmatic evolution of Scotland. *Transactions of the Royal Society of Edinburgh; Earth Sciences* **97**, 213–95.

MACFADYEN, C.C.J. 1992. *The Achanarras Quarry Site of Special Scientific Interest and National Nature Reserve, Caithness*. Earth Science Documentation Series. Scottish Natural Heritage.

MACGREGOR, M. 1916. A Jurassic shoreline. *Transactions of the Geological Society of Glasgow* **16**, 75–85.

McQUILLAN, R., DONATO, J.A. and TULSTRUP, J. 1982. Development of basins in the Inner Moray Firth and the North Sea by crustal extension and dextral displacement of the Great Glen Fault. *Earth and Planetary Science Letters* **60**, 127–139.

MACLENNAN, A.M. and TREWIN, N.H. (1989). Palynofacies and sedimentology of the Late Bathonian–Mid. Callovian in the Inner Moray Firth. *In*: Batten, D.J. (ed.) *Studies in NW European micropalaeontology and palynology*. British Micropalaeontological Society Series, 92–117. Ellis Horwood, Chichester.

MARSHALL, J.E.A., BROWN, J.G. and HINDMARSH, S. 1985. Hydrocarbon source rock potential of the Devonian rocks of the Orcadian Basin, *Scottish Journal of Geology* **21**, 301–320.

MARSHALL, J.E.A., ROGERS, D.A. and WHITELEY, M.J. 1996. Devonian marine incursions into the Orcadian Basin, Scotland. *Journal of the Geological Society, London* **153**, 451–66.

MARSHALL, J.A.E., ASTIN, T.R., BROWN, J.F., MARK-KURIK, E. and LAZAUSKIENE, J. 2007. Recognising the Kacak Event in the Devonian terrestrial environment and its implications for understanding land–sea interactions. *In*: Becker, R.T. and Kirchgasser, W.T. (eds) *Devonian Events and Correlations*. Special Publications, Geological Society, London **279**, 133–55.

MILLER, H. 1841. *The Old Red Sandstone*. John Johnstone, Edinburgh.

MILLER, H. 1854. The fossiliferous deposits of Scotland. *Proceedings of the Royal Physical Society of Edinburgh* **1**, 1–29.

MILLER, H. 1859. *Sketch book of popular geology*. Thomas Constable and Company, Edinburgh.

MURCHISON, R.I. 1827. On the coal-field of Brora in Sutherlandshire. *Transactions of the Geological Society of London* **2**, 293–326.

MYKURA, W. 1976. British Regional Geology, Orkney and Shetland M.S.O. Edinburgh.

MYKURA, W. 1991. Old Red Sandstone. In: Craig, G.Y. (ed.) *Geology of Scotland*, (3rd Edn) 297–344. The Geological Society, London.

NEVES, R. and SELLEY, R.C. 1975. A review of the Jurassic rocks of North-East Scotland. *In*: Finstad, K.G., Selley, R.C. (eds) *Proceedings of the Jurassic Northern North Sea Symposium Stavanger, Sept. 1975*. Norsk Petroleum-forening. JNNSS/5, 1–29.

NEWMAN, M.J. 2002. A new naked jawless vertebrate from the Middle Devonian of Scotland. *Palaeontology* **45**, 933–941.

NEWMAN, M.J. and DEAN, M.T. 2005. *A biostratigraphical framework for geological correlation of the Middle Devonian strata in the Moray–Ness Basin Project area*. British Geological Survey Internal

Report **IR/05/160**.
NEWMAN, M.J. and DEN BLAAUWEN, J.L. 2007. A new dipnoan fish from the Middle Devonian (Eifelian) of Scotland. *Palaeontology* 50, 1403–1419.
NEWMAN, M.J. and DEN BLAAUWEN, J.L. 2008. New information on the enigmatic Devonian vertebrate *Palaeospondylus gunni*. *Scottish Journal of Geology* 44, 89–91.
NEWMAN, M.J. and TREWIN, N.H. 2001. A new jawless vertebrate from the Middle Devonian of Scotland. *Palaeontology*, 44, 43–51.
NEWMAN, M.J. and TREWIN, N.H. 2008. Discovery of the arthrodire genus *Actinolepis* (class Placodermi) in the Middle Devonian of Scotland. *Scottish Journal of Geology* 44, 83–88.
NORTON, M.G., McCLAY, K. and WAY, N.A. 1987. Tectonic evolution of Devonian basins in northern Scotland and southern Norway. *Norsk Geol. Tidsskrift*, 67, 323–38.
OMAND, D. 1973. The glaciation of Caithness. Univ. Strathclyde MSc thesis. (Unpubl.)
OWEN, J.S. 1995. *Coal mining at Brora, Sutherland 1529–1974*. Inverness Highland Libraries.
PANKHURST, R.J. 1982. Geochronological tables for British igneous rocks. *In*: Sutherland, D.S. (ed.) *Igneous rocks of the British Isles*. 575–581. John Wiley & Sons, Chichester.
PANKHURST, R.J. and SUTHERLAND, D.S. 1982. Caledonian granites and diorites of Scotland and Ireland. *In*: Sutherland, D.S. (ed.) *Igneous rocks of the British Isles*. 149–190. John Wiley & Sons, Chichester.
PARNELL, J. 1983. The distribution of hydrocarbon minerals in the Orcadian basin. *Scottish Journal of Geology* 19, 205–231.
PARNELL, J. 1986. Devonian Magadi-type cherts in the Orcadian Basin, Scotland. *Journal of Sedimentary Petrology* 56, 495–500.
PEACOCK, J.D., BERRIDGE, N.G., HARRIS, A.L. and MAY, F. 1968. *The geology of the Elgin district*. Memoirs of the Geological Survey, Scotland.
PEARSON, M.J. and WATKINS, D. 1983. Organofacies and early maturation effects in Upper Jurassic sediments from the Inner Moray Firth Basin, North Sea. *In*: Brooks, J. (ed.) Petroleum Geochemistry and Exploration of Europe. *Special Publications, Geological Society, London* 12, 147–160. Blackwell.
PETERS, K.E., MULDOWAN, J.M., DRISCOLE, A.R. and DEMAISON, G.J. 1989. Origin of Beatrice oil by co-sourcing from Devonian and Middle Jurassic source rocks, Inner Moray Firth, United Kingdom. *Bulletin American Association of Petroleum Geologists* 73, 454–471.
PHEMISTER, J. 1960. *British Regional Geology: The Northern Highlands*. (3rd Edn). Edinburgh, H.M.S.O.
PICKERING, K.T. 1983. Small scale syn-sedimentary faults in the Upper Jurassic 'Boulder Beds'. *Scottish Journal of Geology* 19, 169–181.
PICKERING, K.T. 1984. The Upper Jurassic 'Boulder Beds' and related deposits: a fault-controlled submarine slope, N. E. Scotland. *Journal of the Geological Society, London* 141, 357–74.
PIDGEON, R.T. and AFTALION, M. 1978. Cogenetic and inherited zircon U/Pb systems in granites: Palaeozoic granites of Scotland and England. *In*: Bowes, D.R. and Leake, B.E. (eds) *Crustal evolution in north-western Britain and adjacent regions*. Geological Journal Special Issue 10, 183–220.
PLANT, J. and COLEMAN, R.F. 1972. Application of neutron activation analyses to the evaluation of placer gold concentrations. *In*: Jones, M.J. (ed.) *Geochemical exploration 1972 – Prospecting in glaciated terranes*, 373–87. IMM, London.
RAMSAY, A.C. 1865. The glacial theory of lake basins. *Philosophical Magazine* 29, 285.
RAYNER, E.H. 1963. The Achanarras Limestone of the Middle Old Red Sandstone, Caithness, Scotland. *Proceedings of the Yorkshire Geological Society* 34, 1–44.
READ, H.H. 1931. *The geology of central Sutherland*. Memoirs of the Geological Survey Scotland (Sheets 108, 109).
READ, H.H., RUSS, G. and PHEMISTER, J. 1925. *The geology of the country around Golspie, Sutherland*. Memoirs of the Geological Survey Scotland.
RICE, C.M. 2002. Metalliferous minerals. *In*: Trewin, N.H. (ed.) *The Geology of Scotland* (4th Edn) 431–448. The Geological Society, London.
RICHARDS, P.C., LOTT, G.K., JOHNSON, H., KNOX, R.W.O'B. and RIDING, J.B. 1993. Jurassic of the Central and Northern North Sea. *In*: Knox, W.R.O'B. and Cordey, W.G. (eds) *Lithostratigraphic nomenclature of the UK North Sea*. BGS, Nottingham.
RICHARDSON, J.B. 1967. Some British Lower Devonian spore assemblages and their stratigraphical significance. *Reviews of Palaeobotany and Palynology* 1, 111–29.
RILEY, L.A. 1980. Palynological evidence of an early Portlandian age for the uppermost Helmsdale Boulder Beds, Sutherland. *Scottish Journal of Geology* 16, 29–31.

ROBERTS, A. 1989. Fold and thrust structures in the Kintradwell 'Boulder Beds', Moray Firth. *Scottish Journal of Geology* **25**, 173–86.

ROBERTS, A.M., BADLEY, M.E., PRICE, J.D. and HUCK, I.W. 1990. The structural history of a transtensional basin: Inner Moray Firth, NE Scotland. *Journal of the Geological Society, London* **147**, 87–103.

ROGERS, D.A. and ASTIN, T.R. 1991. Ephemeral lakes, mud-pellet dunes and wind-blown sand and silt: reinterpretations of Devonian lacustrine cycles in north Scotland. *In*: Anadón, P., Cabrera, L. and Kelts, K. (eds) *Lacustrine Facies Analysis*. International Association of Sedimentologists Special Publication **13**, 199–221.

ROGERS, D.A., MARSHALL, J.E.A. and ASTIN, T.R. 1989. Devonian and later movements on the Great Glen fault system, Scotland. *Journal of the Geological Society, London* **146**, 369–72.

SAXON, J. 1975. *The fossil fishes of the North of Scotland* (2nd Edn). 1–49. Caithness Books, Thurso.

SAXON, J. (undated). The Kildonan gold rush. Caithness Books, Thurso.

SEWARD, A.C. 1911. The Jurassic flora of Sutherland. *Transactions of the Royal Society of Edinburgh* **47**, 643–709.

SISSONS, J.B. 1968. *The evolution of Scotland's scenery*. Oliver and Boyd, Edinburgh.

STEPHEN, K.J., UNDERHILL, J.R., PARTINGTON, M.A. and HEDLEY, R.J. 1993. The genetic sequence stratigraphy of the Hettangian to Oxfordian succession, Inner Moray Firth. *In*: Parker, J.R. (ed.) *Petroleum Geology of Northwest Europe: Proceedings of the 4th Conference*. Geological Society, London.

STEVENS, V. 1991. The Beatrice Field, Block 11/30a, UK North Sea. *In*: Abbotts, I.L. (ed.) *United Kingdom Oil and Gas Fields, 25 Years Commemorative Volume*. Geological Society Memoir **14**, 245–52.

STOPES, M.C. 1907. The flora of the Inferior Oolite of Brora. *Quarterly Journal of the Geological Society of London* **63**, 375–82.

STOW, D.AV., BISHOP, C.D. and MILLS, S.J. 1982. Sedimentology of the Brae oilfield, North Sea: fan models and controls. *Journal of Petroleum Geology* **5**, 129–48.

STRACHAN, R.A., FRIEND, C., ALSOP, G.I. and MILLER, S. (Eds) 2009a (in press). *A Field Guide to the Moine Geology of the NW Highlands of Scotland*. Edinburgh and Glasgow Geological Societies.

STRACHAN, R.A., HOLDSWORTH, R.E., and ALSOP, G.I. 2009b (in press). Introduction to the Moine Geology of the NW Highlands of Scotland. *In*: Strachan *et al.* (eds) *A Field Guide to the Moine Geology of the NW Highlands of Scotland*. Edinburgh and Glasgow Geological Societies.

STRACHAN, R.A., HOLDSWORTH, R.E., FRIEND, C., BURNS, I. and ALSOP, G.I. 2009c (in press). Excursion 13 North Sutherland. *In*: Strachan *et al.* (eds) *A Field Guide to the Moine Geology of the NW Highlands of Scotland*. Edinburgh and Glasgow Geological Societies.

STRACHAN, R.A., SMITH, M., HARRIS, A.L. and FETTES, D.J. 2002. The Northern Highland and Grampian terranes. *In*: Trewin, N.H. (ed.) *The Geology of Scotland* (4th Edn) 81–147. The Geological Society, London.

SYKES, R.M. 1975a. The stratigraphy of the Callovian and Oxfordian stages (Middle–Upper Jurassic) in Northern Scotland. *Scottish Journal of Geology* **11**, 51–78.

SYKES, R.M. 1975b. Facies and faunal analysis of the Callovian and Oxfordian Stages (Middle-Upper Jurassic) in N. Scotland and E. Greenland. Univ. Oxford Ph.D. Thesis (unpubl.).

TAIT, D. 1912. On a large glacially transported mass of Lower Cretaceous rock at Leavad in the County of Caithness. *Transactions of the Edinburgh Geological Society* **10**, 1–9.

TAYLOR, J.C.M. 1990. Upper Permian–Zechstein. *In*: Glennie, K.W. (ed.) *Introduction to the Petroleum geology of the North Sea* (3rd Edn). 153–190. Blackwells.

THOMSON, K. and UNDERHILL, R. 1993. Controls on the development and evolution of structural styles in the Inner Moray Firth Basin. *In*: Parker, J.R. (ed.) *Petroleum Geology of Northwest Europe: Proceedings of the 4th Conference* Geological Society, London. 1167–78.

THOMSON, K.S., SUTTON, M. and THOMAS, B. 2003. A larval Devonian lungfish. *Nature*, **426**, 833–4.

TREWIN, N.H. 1985. Mass mortalities of Devonian fish – the Achanarras fish bed, Caithness, *Geology Today*, **1**, 45–9.

TREWIN, N.H. 1986. Palaeoecology and sedimentology of the Achanarras fish bed of the Middle Old Red Sandstone, Scotland. *Transactions of the Royal Society of Edinburgh: Earth Sciences* **77**, 21–46.

TREWIN, N.H. 1987. Pennan, unconformity within the Old Red Sandstone. *In*: Trewin, N.H., Kneller, B.C. and Gillen, C. (eds) *Excursion guide to the geology of the Aberdeen area*. 127–130. Scottish Academic Press.

TREWIN, N.H. 1989. The petroleum potential of the Old Red Sandstone of northern Scotland. *Scottish Journal of Geology* **25**, 201–25.

TREWIN, N.H. 1991. *Jurassic sedimentation and tectonics in the Brora–Helmsdale area, and Old Red Sandstone fluvial and lacustrine facies in N. Scotland.* Field Guide No. 8. Compiled for the 13th International Sedimentological Congress, Nottingham, 1990. 1–86. British Sedimentological Research Group, Cambridge.

TREWIN, N.H. 1992. 'Subaqueous shrinkage cracks' in the Devonian of Scotland re-interpreted – Discussion. *Journal of Sedimentary Petrology* **62**, 921–2.

TREWIN, N.H. (ed.) 2002. *The Geology of Scotland* (4th Edn) The Geological Society, London.

TREWIN, N.H. 2008. *Fossils Alive! or new walks in an old field.* Dunedin Academic Press, Edinburgh.

TREWIN, N.H. and KNELLER, B.C. 1987. Old Red Sandstone and Dalradian of Gamrie Bay. *In*: Trewin, N.H., Kneller, B.C. and Gillen, C. (eds) *Excursion guide to the geology of the Aberdeen area.* 113–126. Scottish Academic Press.

TREWIN, N.H. and ROLLIN, K.E. 2002. Geological history and structure of Scotland. *In*: Trewin, N.H. (ed.) *The Geology of Scotland* (4th Edn), 1–26. The Geological Society, London.

TREWIN, N.H. and THIRLWALL, M.F. 2002. The Old Red Sandstone. *In*: Trewin, N.H. (ed.) *The Geology of Scotland* (4th Edn), 213–250. The Geological Society, London.

TURNER, C.C., COHEN, J.M., CONNELL, E.R. and COOPER, D.M. 1987. A depositional model for the South Brae Oilfield. *In*: Brooks, J. and Glennie, K.W. (eds.) *Petroleum Geology of North-west Europe.* Proceedings 3rd Conf. Pet. Geol. N.W. Europe. 853–864. Graham and Trotman, London.

TWEEDIE, J.R. 1979. Origin of uranium and other metal enrichments in the Helmsdale Granite, eastern Sutherland, Scotland. *Transactions of the Institute of Mining and Metallurgy* **88**, B145–53.

TWEEDIE, J.R. 1981. The origin of uranium and other metal concentrations in the Helmsdale Granite and the Devonian sediments of the north-east of Scotland. Univ. Aberdeen PhD Thesis. (unpubl).

UNDERHILL, J.R. 1991. Implications of Mesozoic–Recent basin development in the western Inner Moray Firth, UK. *Marine and Petroleum Geology* **8**, 359–69.

UNDERHILL, J.R. 1994. Discussion on thepaleoecology and sedimentology across a Jurassic fault scarp, NE Scotland. *Journal of the Geological Society, London* **151**, 729–31.

UNDERHILL, J.R. 1998. Jurassic. *In*: Glennie, K. W. (ed.) *Petroleum Geology of the North Sea* 4th Edn. Blackwell, Oxford. 245–293.

UNDERHILL, J.R. and BRODIE, J.A. 1993. Structural geology of Easter Ross, Scotland: implications for movement on the Great Glen Fault. *Journal of the Geological Society, London* **150**, 515–27.

UNDERHILL, J.R. and PARTINGTON, M.A. 1993. Jurassic thermal doming and deflation in the North Sea: implications of the sequence stratigraphic evidence. *In*: Parker, J. R. (ed.) *Petroleum Geology of Northwest Europe: Proceedings of the 4th Conference.* Geological Society, London. 337–346.

VAGLE, G.B., HURST, A., DYPVIK, H. 1994. Origin of quartz cements from the Jurassic of the Inner Moray Firth (UK). *Sedimentology* **41**, 363–77.

VAGLE, G.B., HURST, A., DYPVIK, H. 1995. Origin of quartz cements from the Jurassic of the Inner Moray Firth (UK)Discussion. *Sedimentology* **42**, 375–8.

VAN DER BURGH, J. 1987. Macroflora of the Kimmeridgian of Sutherland – a preliminary report. *Documenta Naturae* **41**, 1–10. Munchen.

VAN DER BURGH, J. and VAN KONIJNENBURG-VAN CITTERT, J.H.A. 1984. A drifted flora from the Kimmeridgian (Upper Jurassic) of Lothbeg Point, Sutherland, Scotland. *Reviews of Palaeobotany and Palynology* **43**, 359–96.

WALKER, E.F. 1985. Arthropod ichnofauna of the Old Red Sandstone at Dunure and Montrose, Scotland. *Transactions of the Royal Society of Edinburgh: Earth Sciences* **76**, 287–297.

WATERSTON, C.D. 1951. The stratigraphy and palaeontology of the Jurassic rocks of Eathie (Cromarty). *Transactions of the Royal Society of Edinburgh* **62**, 33–51.

WESTOLL, T.S. 1951. The vertebrate-bearing strata of Scotland. *Report 18th International Geological Congress* **2**, 5–21.

WIGNALL, P.B. and PICKERING, K.T. 1993. Palaeoecology and sedimentology across a Jurassic fault scarp. N E Scotland. *Journal of the Geological Society, London* **150**, 323–340.

WILSON, G.V., EDWARDS, W., KNOX, J., JONES, R.C.B., and STEPHENS, J.B. 1935. *The Geology of the Orkneys.* Memoir of the Geological Survey of Great Britain.

WINDLE, T.M.F. 1979. Reworked Carboniferous spores: an example from the Lower Lias of northeast Scotland. *Reviews of Palaeobotany and Palynology* **27**, 173–84.

WOODWARD, H.B. 1911. *In*: Seward, A.C. The Jurassic flora of Sutherland. *Transactions of the Royal Society of Edinburgh* **47**, 643–709.

ZIEGLER, P.A. 1982. *Geological atlas of western and central Europe.* Shell Internationale Petroleum Maatschappij 130pp, The Hague.

The Bridge Hotel

Dunrobin Street, Helmsdale, KW8 6 JA, Sutherland, Scotland
Tel: 01431 821 100
www.bridgehotel.net

The Bridge Hotel rests by the banks of the famous Helmsdale River, near a working harbour bobbing with fishing boats and below a Strath where hundreds of deer gently roam the wild outdoors. Built circa 1816 as a traveller's inn (Ross's Commercial Inn) and situated in the heart of the village of Helmsdale on the East Coast of Scotland, the building is steeped in history and tradition, positioned at the head of the old Helmsdale River Bridge. Served by two airports – Inverness and Wick – it is conveniently situated on the main axis between Inverness, Thurso and John o' Groats, offering a perfect base for exploring the wonders of the Highlands.

'The Green Stag' at the Bridge Hotel in Helmsdale pays homage to Scotland's best produce, straight from the hills and the sea and believes in keeping things simple and unembellished, both in terms of hospitality and the food that is served up

ORCADIAN STONE COMPANY

MAIN STREET, GOLSPIE, SUTHERLAND, KW10 6RH
Telephone: Golspie 01408 633483
e-mail: orcadianstone@aol.com
Visit our website at www.orcadianstone.co.uk

**Natural Stone Products   Gemstones   Mineral Specimens
Geological Exhibition   Shop**

The exhibition and shop are open from Easter to the end of October,
Monday to Saturday, 9.00am to 5.30pm.
The shop also opens for three weeks before Christmas.
There is an admission charge to the exhibition of £4 for adults and £1 for Children.
University and school parties are free.
The exhibition is not suitable for children under 8 years.